Nanomaterials for Applications in Energy and Biology 2018

Topical Conference at the 2018 AIChE Annual Meeting

Pittsburgh, Pennsylvania, USA
28 October - 2 November 2018

ISBN: 978-1-5108-7638-5

Printed from e-media with permission by:

Curran Associates, Inc.
57 Morehouse Lane
Red Hook, NY 12571

Printed by Curran Associates, Inc. (2019)

For permission requests, please contact AIChE
at the address below.

AIChE
120 Wall Street, FL 23
New York, NY 10005-4020

Phone: (800) 242-4363
Fax: (203) 775-5177

www.aiche.org

Additional copies of this publication are available from:

Curran Associates, Inc.
57 Morehouse Lane
Red Hook, NY 12571 USA
Phone: 845-758-0400
Fax: 845-758-2633
Email: curran@proceedings.com
Web: www.proceedings.com

TABLE OF CONTENTS

Some abstracts in this book refer to images that were not made available by the publisher

520989 Selectivity of Photocatalytic Conversion of Carbon Dioxide Modulated By Surface Ligands on Cu_2O/TiO_2 Particles

Abstract:

Conversion of CO_2 into CO and hydrocarbon using solar irradiation has been intensively studied, yet efficiency and selectivity of CO_2 photoconversion remains to be improved. In photocatalysis, CO_2 and CO, which is intermediate state of CO2 reduction, adsorbed on the catalyst surface plays an important role in determining activity and selectivity because stabilized CO_2 and CO can react with adsorbed proton or hydroxyl groups without detachment. In this study, we carried out gas-phase conversion of CO_2 using Cu/TiO_2 nanoparticles as photocatalyst, showing significant enhancement when ligand is passivated on the surface. Interestingly, CH_4 production rate increased 2 times and CO production rate decreased when taurine passivated on Cu/TiO_2 photocatalyst. However, CO production rate increased when ethylenediamine treated. These results proved by in-situ Fourier-transform infrared spectroscopy (FT-IR) under photocatalytic CO_2 conversion condition. In addition, density functional theory (DFT) calculation shows surface ligand can control the selectivity of CO_2 conversion by changing of CO binding strength. These findings provide a platform for the design of surface chemistry on photocatalytic conversion of CO_2.

From this work, I demonstrate control for the selectivity of CO_2 conversion in photocatalyst using taurine and ethylenediamine ligands treatment. The taurine treatment enhance the CO binding strength of Cu_2O surface, so the CH_4 selectivity is 2 times enhanced. However, the CO selectivity is dramatically enhanced when ethylenediamine treated. These changes attributed to the control of CO and COOH binding strength of Cu_2O surfaces. Therefore, we investigate about the correlation between the surface interaction of the ligands and the selectivity of CO_2 photocatalysts. This work is proved by situ DRIFTS, photoactivity measurements and DFT calculation method. These DFT calculation results also proved by photocatalytic reaction and DRIFTS measurements. Also, the important finding was that CO_2^-, which is primary CO_2 reduction state, changed to different intermediate state. If the strong CO binding condition, the CO_2^- would be changed to bidentate or polydentate carbonate form in taurine passivation, but if the weak CO binding condition, the CO_2^- could converted to CO gas immediately in ethylenediamine passivation. This study has provided new information to correlate the surface chemistry with CO_2 reduction activity by Cu_2O/TiO_2. The ligand passivation appears to be a promising method for CO_2 reduction, and therefore we can control the activity and selectivity of CO_2 reduction.

522628 Oxygen-Deficient Monoclinic Tungsten Oxide Nanowires for Spectrally Selective Electrochromic Smart Windows

Abstract:

Dual-band electrochromic smart windows where the near-infrared (NIR) and visible (VIS) light transmittance can be dynamically and independently varied can significantly reduce the building energy consumption on heating, ventilation, air conditioning (HVAC) and lighting.[1] Most of the dual-band electrochromic materials reported to date are composites[1-2] formed by the deliberate and delicate integration of NIR-selective and VIS-modulating components which operate in different potential windows. The use of single-component materials with a good dual-band performance would have been preferred but few have been identified.

We discovered that that the dynamic and independent tuning of NIR and VIS transmission is possible with oxygen-deficient monoclinic tungsten oxide nanowires (m-WO_{3-x} NWs) - a single-component material.[3] Its NIR transmittance can be modulated by localized surface plasmon resonance (LSPR) in the 4-2.8 V window; and by LSPR and phase-transition (from dielectric to metallic) in the 2.8-2.6 V window. On the other hand, its VIS light transmittance is modulated by bandgap transitions (intraband and interband transitions) in the lower 2.6-2 V window. Consequently cast films of the m-WO_{3-x} NWs can operate in three distinct modes: the bright mode (4 V, NIR and VIS transparent), the cool mode (2.8-2.6 V, mostly NIR blocking) and the dark mode (2 V, NIR and VIS opaque). In particular a m-WO_{3-x} NW film in the cool mode (2.6V) can block 92.2% of the NIR to reduce the building solar heat gain, while maintaining a high VIS transmittance of 64.6% for daylighting. The cast films of m-WO_{3-x} NWs also provide a high optical modulation of the full solar spectrum (91.7%, 92.7%, 94.6%, 87.3% at 633, 800, 1200, 1600 nm respectively), a high coloration efficiency (101.7 and 184.3 cm^2 C^{-1} at 633 and 1200 nm respectively) and good cycle stability (8.3% of optical loss at 633 nm after 1000 cycles).

References

[1] A. Llordés, G. Garcia, J. Gazquez and D. J. Milliron, *Nature*, **500**, 323 (2013).

[2] A. Llordes, Y. Wang, A. Fernandez-Martinez, P. Xiao, T. Lee, A. Poulain, O. Zandi, C. A. Saez Cabezas, G. Henkelman and D. J. Milliron, *Nat. Mater.*, **15**, 1267 (2016).

[3] S. Zhang, S. Cao, T. Zhang, Q. Yao, A. Fisher, J. Y. Lee, *Mater. Horiz.*, **5**, 291 (2018)

531639 Solar-Driven Photocatalytic Reforming of Glycerol for Hydrogen Production over Ternary Cu/THS/Graphene Photocatalyst: Effects of Reaction Conditions

Abstract:

Solar photocatalytic hydrogen production from glycerol reforming over a ternary 0.5wt.%Cu/THS/graphene$_{4wt.\%}$ photocatalyst was studied under various reaction conditions. The influence of reaction parameters such as initial glycerol concentration, solution pH, and photocatalyst dosage on the photocatalytic hydrogen evolution activity of the ternary 0.5wt.%Cu/THS/graphene$_{4wt.\%}$ catalyst was investigated. It was found that all of the above mentioned parameters influenced the photocatalytic activity of the ternary catalyst. The best operating initial glycerol was found to be 10 % (v/v); the maximum photocatalytic activity was achieved when the solution was slightly acidic (pH 6); and the catalyst dosage of 80mg achieved the highest rate of photocatalytic hydrogen evolution.

532040 Structural Effect Study in an Assembled Nano-Heterojunction Towards Designing a Visible Light Photocatalyst for H_2 Generation

Abstract:

Hydrogen production using photocatalytic water splitting has drawn great interest in our scientific community due to the demand for renewable energy. In photocatalytic water splitting, heterojunction formation of different functional materials in a single photocatalyst shows potential advantages over a single material photocatalyst. Advantages include extended solar light absorption from UV to visible light range, reduces photoexcited electron-hole recombination by enhancing charge separation, and increases the photo corrosion resistibility. In designing the assembled nano-heterojunction, a UV- active perovskite material ($SrTiO_3$) of different structures was decorated with several functional materials including visible light active (CdS), plasmonic metal (Au), and a cocatalyst. Sol gel assisted electrospinning and hydrothermal methods were used to synthesize the nano-heterojunctions. These nano-heterojunctions were characterized by field emission scanning electron microscopic with energy dispersive spectroscopic observations, X-ray diffraction patterns, N_2 gas absorption-desorption isotherm measurements, and UV-visible spectrophotometry. Structural effect of $SrTiO_3$ and CdS in the assembled nano-heterojunctions were investigated in terms of charge recombination and H_2 generation from photocatalytic water splitting. Charge recombination phenomena were studied by photoluminescence spectra and photocurrent measurements. High photocatalytic activities of these assembled nano-heterojunctions are expected due to multifunctional materials assembly, preferential heterostructure formation, and enhanced visible light absorption.

533865 Cooperative Effect of Co and Zr Co-Doping on the Photoelectrochemical Water Splitting Performance of Hematite

Abstract:

The energy crisis results in the increased demand for clean fuels such as hydrogen energy. One of the most promising methods to produce hydrogen is photoelectrochemical water splitting, which realizes the conversion from solar energy to chemical energy that is stored in hydrogen molecules. Photoanodes play an important role in the photoelectrochemical water splitting process.

Hematite (Fe_2O_3) is considered as a promising photoanode material due to its stability, abundance and suitable band gap for light absorption. However, its intrinsic defects, such as poor conductivity, suppress the water splitting performance. Doping is a useful way to improve the conductivity of hematite. In this work, Co and Zr are incorporated into iron oxide through in-situ synthesis. Co doping exhibits a positive effect on the hematite photoelectrocatalytic activity, while Zr doping lowers the activity slightly. Furthermore, the cobalt and zirconium co-doped Fe_2O_3 photoanode shows a better performance for photoelectrochemical water oxidation than that of Co doped photoanode. The photocurrent density of co-doped Fe_2O_3 reaches 0.55 mA cm^{-2} at 0.23 V (vs. AgCl), which is increased by 27.9% compared with the raw Fe_2O_3. This result reveals the cooperative effect between Co and Zr on improving the Fe_2O_3 performance towards water oxidation.

516552 Photo-Catalytic Degradation of Pharmaceuticals in Water Matrix Under Simulated Solar Light Using BiOCl/BiOI

Abstract:

BiOCl/BiOI heterojunction catalyst was rapidly prepared at room temperature using microwave assisted method. Photo-catalyst experiment performed in degrading Orange G reveals that BiOCI/BiOI composite with molar ratio of 0.85 to 0.15 exhibited the highest photo-activity in both simulated solar and LED visible light irradiation. Result shows that BiOCl/BiOI composite was better than TiO_2 (P25) and the individual catalyst components. The optimum BiOCl/BIOI composite was further used to degrade pharmaceuticals (Carbamazepine, Ibuprofen, Sulfamethoxazole, and Diclofenac) in water matrix. The high catalyst activity experienced by $BiOCl_{0.85}/BiOI_{0.15}$ is attributed to reduction in electron-hole recombination rate, high surface area for adsorption and harnessing of visible light photons by BiOI in the composite. In addition, a novel photocatalytic continuous flow reactor with large immobilized catalyst area was designed and fabricated for degrading Emerging Contaminants (Pharmaceuticals) in water. The results obtained from these research work will both proffer an energy efficient method in preparing highly active photo-catalyst and a cost effective technique for degrading contaminants in water.

527144 Invited Talk: Nanotherapeutics for Neuroprotection in the Developing Brain

Abstract:

BACKGROUND: Neonatal brain injury is a major cause of morbidity and mortality in neonates. No therapy is currently curative. Therefore, more effective neuroprotective therapies are needed to improve neurological outcomes in newborns with brain injury. We aim to develop an effective neuroprotective treatment using enzyme or hydrophobic drug-loaded nanoparticles in the injured neonatal brain. We explore the therapeutic benefit of superoxide dismutase (SOD), catalase, and curcumin, broad-acting drugs that have antioxidant, anti-inflammatory, and anti-apoptotic effects, but low aqueous solubility, low stability, low bioavailability, and rapid first-pass hepatic metabolism. Polymeric nanoparticles are a useful tool to overcome these limitations, and can provide high drug encapsulation, sustained release, improved drug solubility and circulation kinetics.

METHODS: Protein or drug-loaded poly(lactic-*co*-glycolic acid) poly(ethylene glycol) (PLGA-PEG) nanoparticles were made, characterized, and evaluated for efficacy in the injured neonatal brain. SOD and catalase nanoparticles were applied to excitotoxic organotypic brain slices. Cell death and mRNA profiles following treatment were measured. Curcumin-PLGA-PEG particles (10 mg/kg curcumin) were administered intraperitoneally (i.p.) at 0.5h, 24h, and 48h after unilateral hypoxic ischemic injury in postnatal 7 (P7) rat pups. Control groups were saline, free curcumin (10 mg/kg dose), and blank nanoparticles. Brain, serum, and organs of interest were evaluated at 72h after HI for global pathology, histopathology, and neuropathology.

RESULTS: In neonatal brain injury models, PLGA-PEG nanoparticles uptake in the hippocampus, striatum, thalamus, and cortex, all regions of injury in the neonatal brain. SOD nanoparticles were able to halt excitotoxicity processes, decreasing oxidative stress and ongoing inflammation in injured brain slices compared to free drug controls. SOD outperformed catalase nanoparticles in this model. In HI pups, curcumin-PLGA-PEG nanoparticles reduce global and neuronal injury in the HI brain three days after systemic administration compared to all controls. We see no toxicity in the liver or kidney of curcumin-PLGA-PEG treated pups, and no particle uptake in healthy controls.

CONCLUSIONS: Strategies that provide better neuroprotection in the neonatal period can address the critical need for effective therapeutics for brain injury. We demonstrate the potential of nanotechnology-based neuroprotective therapies for neonatal brain injury, an area largely underserved by novel drug delivery technologies.

536662 In Vivo Imaging of Larval Zebrafish Neurochemistry with Near-Infrared Dopamine Nanosensors

Abstract:

Elucidating the relationship between the chemical activity and neuron connectivity to higher order function, such as behavior and cognition, is a fundamental goal in neuroscience. Although numerous tools exist to measure the electrical activity of neurons, the ability to directly monitor chemical communication between cells via the release and uptake of neurotransmitters and neuromodulators, such as dopamine, remains an unmet challenge. Recently, developments in the non-covalent functionalization of single-walled carbon nanotubes has produced a new class of fluorescent nanosensors for imaging dopamine. These dopamine nanosensors are water-soluble, non-toxic, and emit fluorescence in the near-infrared (NIR) window without photobleaching, making them ideally suited for long-term fluorescence imaging in high-scattering tissues like the brain. Additionally, their size and binding kinetics have been shown to capture the relevant spatial (um) and temporal (ms) dynamics of neurotransmission in the brain extracellular space.[1] Towards the goal of achieving real-time imaging of dopaminergic neuromodulation in awake and behaving animals, we present a NIR wide-field fluorescence imaging system compatible with SWNT-based dopamine nanosensors. The system leverages continuous wave epifluorescent Koehler Illumination to image live specimens. To demonstrate in vivo imaging capabilities, SWNT dopamine nanosensors injected into the brain of larval zebrafish (Danio rerio, 4-7 dpf) are shown to localize in the extracellular matrix in different subregions of the brain, where dopaminergic synapses are present. Additionally, genetically encoded calcium indicators confirm the nanosensors minimally impact general health in their presence. Our results demonstrate the utility of infrared nanosensors to visualize dopamine release in the brain of an awake animal towards a better understanding of in vivo neuromodulatory dynamics.

1 A. G. Beyene, I. R. McFarlane, R. L. Pinals and M. P. Landry, *ACS Chem. Neurosci.*, 2017, **8**, 2275â€"2289.

540422 Functional Mesoporous Silica for Immunoengineering and Immunotherapy

Abstract:

Failures of the native immune system to eradicate tumors are mostly due to the immunosuppressive environments created by the tumor. In order to generate robust and durable antitumor adaptive immune responses, it is crucial to overcome the immunosuppressive environments that affect to the immune cells. In this presentation I will discuss on the design and application of functional mesoporous silica to cancer vaccine by activating immune cells. The functional mesoporous silica includes 3D scaffold composed of injectable mesoporous microparticles, hierarchically porous scaffold, and large-pore mesoporous silica nanoparticles to activate dendritic cells and cytotoxic T cells in vivo. The functional mesoporous silica could act as delivery vehicles of cancer antigen and adjuvant, and as building blocks to form cellular microenvironment in the body for the manipulation of host dendritic cells. The appropriate design and engineering of functional nanobiomaterials could have the potential to treat cancer and inflammatory disease in the future.

515049 Green Synthesis of Fluorescent Nanomaterials for Optical Bioimaging and Beyond

Abstract:

Fluorescent nanoparticles have received intense scientific attention and offer promising applications in many fields from optoelectronic devices to cancer treatment. The development of robust mass-scale nanoparticle synthesis methods is one of key materials challenges in moving towards fluorescent nanoparticles for commercialization and advanced applications. In this talk, I will summarize our work on green and scalable synthesis of fluorescent nanoparticles, including biomass-derived carbon dots, subcritical water technology for nanonization of organic compounds and preparation of inorganic nanoparticles by high-gravity process intensification technology. The applications of related fluorescent nanoparticles in biomedical imaging and therapy from cells to animals will be introduced. A brief overview of this exciting field, along with some challenges and opportunities, will also be presented.

References

[1] Z. Wang, Y. Pu, D. Wang, J. Shi, J.-X. Wang, J.-F. Chen, *AIChE J.*, 2018, 64, 1330.

[2] Y. Pu, F. Cai, D. Wang, J.-X. Wang, J.-F. Chen, *Ind. Eng. Chem. Res.*, 2018, 57, 1790.

[3] J. Leng, J. Chen, D. Wang, J.-X.Wang, Y. Pu, J.-F. Chen, *Ind. Eng. Chem. Res.*, 2017, 56, 7977.

[4] D. Wang, L. Zhu, Y. Pu, J.-X. Wang, J.-F. Chen, L. Dai, *Nanoscale*, 2017, 9, 11214

[5] D. Wang, L. Zhu, J.-F. Chen, L. Dai, *Angew. Chem. Int. Ed.*, 2016, 55, 10795.

531062 Biological Self-Assembly and Recognition Used to Synthesize and Guide Next Generation of Hybrid Bio-Nano-Materials

Abstract:

Free-standing, high aspect ratio sulfur-doped carbon nanodots-based hybrid nanowires with a microtubular aspect were synthesized using self-recognition and self-assembly processes of tubulin, a biological molecule precursor of cytoskeletal cellular and structural filaments. Physico-chemical characterizations (e.g., morphology, diameter, spectra absorbance etc.) of the user-synthesized hybrid bio-nanowires were performed using a variety of morphological (transmission and electron microscopy, and atomic force microscopy) and spectroscopical (X-ray photoelectron and energy dispersive spectroscopy) techniques. Biofunctionality was demonstrated by mimicking cellular transport based on kinesin, a motor protein capable to recognize, bind onto and move such tubulin-based filaments, as well as through both protein circular dichroism and absorbance. Our results indicate that user-synthesized hybrid bio-nanowires could be manipulated *in vitro* under constant chemical energy of adenosine triphosphate (ATP) and have the potential to be implemented in the next generation of synthetic applications. Specifically, our results demonstrate the ability to implement biological function (through the use of tubulin self- assembly and kinesin recognition and bipedal moving) to a non-biological material (sulfur-doped carbon nanodots) for creating the next generation of hybrids that could be manipulated and user-placed for synthetic applications from biosensing to electronic devices.

540040 A Model-based Analysis of the Tissue-Targeting Efficacy of Ligand-Directed Nanoparticles

Abstract:

Active delivery of drug nanoparticles involve modifying the particle surface with a targeting ligand. The rationale is that the surface-attached ligand can direct an intravenously-injected particle to certain receptor proteins overexpressed by cancer cells of the target tissue. However, despite the growing effort to conjugate nanoparticles with a variety of targeting ligands, the therapeutic benefit from such strategy remains poorly investigated. In this study, we develop a Brownian Dynamics model to quantitatively understand the gain in the targeting efficacy arising from such ligand targeting. Using the model, we consider different levels of receptor expression in the target cell and evaluate how that affect the targeting efficacy. For each ligand expression level, we consider two different scenarios: 1) particles are captured only through cell-surface receptor molecules in the absence of non-specific recognition; 2) particle capture occurs through both receptor-mediated and nonspecific membrane recognition. Our analysis suggest that there could only marginal gain in the targeting efficacy from the ligand conjugation. In addition, we show that this gain becomes further small in the presence of nonspecific recognition.

538801 Pendant HDAC Inhibitor SAHA Derivatized Polymer As a Novel Prodrug Micellar Carrier for Anticancer Drugs

Abstract:

Suberoylanilide hydroxamic acid (SAHA), a histone deacetylase inhibitor (HDACI) approved by FDA for the treatment of cutaneous T cell lymphoma, is a promising anticancer drug for various cancers with a unique mode of action. However, it demonstrates limited clinical benefits in solid tumors as a single drug. In order to achieve enhanced and synergistic co-delivery of SAHA and Doxorubicin (DOX), a cleavable SAHA-based prodrug polymer (POEG-*b*-PSAHA), consisting of hydrophilic poly(oligo(ethylene glycol) methacrylate) (POEG) blocks and hydrophobic SAHA segments, has been developed. POEG-*b*-PSAHA prodrug polymer was able to form spherical micelles with a diameter around 60 nm, and well retained the pharmacological activity of SAHA in either inhibiting the proliferation of tumor cells or inducing histone acetylation. DOX formulated in POEG-*b*-PSAHA-based micelles showed a sustained release profile. DOX-loaded POEG-*b*-PSAHA exhibited more potent cytotoxicity towards tumor cells than free DOX and DOX loaded in a pharmacologically â€œinertâ€ nanocarrier, POEG-*b*-POM. Consistently, DOX/POEG-*b*-PSAHA formulation resulted in an improved therapeutic effect *in vivo* compared to free DOX, Doxil, or DOX formulated in POEG-*b*-POM micelles. These results suggest that SAHA-based prodrug micelles may serve as a dual functional carrier for combination strategies in epigenetic-oriented anticancer therapy.

531093 Iron Sulfide Supraparticles As Artificial Viruses for Gene and Gene Editing Therapies

Abstract:

Gene and gene editing therapies have been widely investigated for treatment of inherited or acquired genetic diseases. Efficient delivery of therapeutic agents has become a significant barrier in clinical applications due to the toxicity and instability of the vectors in the complex intracellular environment. Among non-viral vectors, individual inorganic nanoparticles (NPs) have become a popular strategy for nucleic acid delivery. However, the nanoshell geometry of viruses is advantageous for the gene/CRISP cargo protection. Therefore, we synthetized L-cysteine stabilized iron-based inorganic nanoparticles which self-assemble into supraparticles with nanoshell geometry. Transmission electron microscopy (TEM), TEM tomography and dynamic light scattering (DLS) were used to characterize the virus-like supraparticles size, shape, and charge. Our results indicate that virus-like supraparticles contain continuous compartments, are positively charged (25±7.2 mV) and 74±21 nm in diameter. We loaded the DNA in the compartments during the formation of supraparticles. We tested these complexes in circular dichroism, UV-Vis spectroscopy, electrophoretic mobility shift and protection assays. Since iron sulfide is a natural material, it presumably has low cytotoxicity and high biocompatibility. Supraparticles can condense DNA, protect it against degradation, penetrate through cellular membranes and facilitate endolysosomal escape in gene therapy. Therefore, development of these particles can be used as an effective cargo delivery tool for gene and gene editing therapies.

540773 2018 Outlook for Energy: A View to 2040

Abstract:

The Outlook for Energy is ExxonMobil's global view of energy demand and supply through 2040. It also highlights the dual challenge of providing affordable energy to support prosperity while reducing environmental impacts – including the risks of climate change. We use the Outlook to help inform our long-term business strategies and investment plans.
The Outlook anticipates significant changes through 2040 across the world to boost living standards, reshape the use of energy, broaden access to abundant energy supplies, and accelerate decarbonization of the world's energy system to address the risks of climate change.

- Global energy needs will rise about 25 percent over the period to 2040
- Electricity demand will nearly double in non-OECD countries
- Worldwide electricity from solar and wind will increase about 400 percent
- Natural gas will expand its role, led by growth for electricity generation
- Growth in oil demand will be driven by commercial transportation and chemical needs
- Efficiency gains and growing use of less-carbon intensive energy sources will contribute to a nearly 45 percent decline in the carbon intensity of global GDP

Energy is fundamental to modern life, and as the world's population approaches 9 billion people in 2040, we are challenged to improve living standards everywhere. Understanding the factors that drive the world's energy needs – and likely choices to meet those needs – is the mission of the Outlook. By sharing the Outlook with the public, we hope to broaden that understanding. Energy matters to everyone, and we all play a role in shaping its future.

541543 Energy Decarbonisation Scenarios

Abstract:

With the energy sector being associated with 2/3 of the annual greenhouse gas emissions, decarbonizing the energy is a key step to reduce their level. The National Determined Contributions (NDC's) announced so far by the participating countries do not meet the objectives of the Paris-climate agreement, as far as reaching a peak in emissions in the near future or decreasing to a level that is compatible with the stated long-term targets.

Using different scenarios for energy demand and supply, options for energy sector decarbonisation are evaluated. Meeting the 2-degree target requires a major transformation in the power sector, efficiency improvements – specially in the buildings and in the industry, and a revolution in transport towards fuel and modal shifts. With a significant increase in renewable energy, system integration and optimization will also be key in reducing the cost of the energy transition. Meeting more ambitious climate goals ("well below 2 degrees") will push further the technological innovation and implementation requirements throughout the entire energy value chain. A scenario aiming at zero net-emissions by 2060 is presented, along with a comparison to the 2-degree case.

542625 Fundamental Research Needs to Advance Energy Technologies

Abstract:

The U. S. Department of Energy (DOE), Office of Science, Office of Basic Energy Sciences (BES) supports fundamental research in chemical and materials sciences to provide the foundations for new energy technologies and to support DOE missions in energy, environment, and national security. This presentation will discuss opportunities for fundamental research to impact DOE's energy mission "to catalyze the timely, material, and efficient transformation of the nation's energy system and secure U.S. leadership in energy technologies" with a focus on the way we generate, store and use energy nationally. I will provide an overview of BES strategic planning over the past decade that identified priority research directions for advancing energy applications, highlight key scientific advances in these areas, and discuss some future opportunities for modern science, particularly chemical sciences, to accelerate the transformation of the U. S. energy portfolio.

540420 Enzyme-Based Antimicrobial Nanoconjugates

Abstract:

Bacterial infections remain a major public health concern. However, broadspectrum antibiotics largely target redundant mechanisms of bacterial survival and lead to gained resistance owing to microbial evolution. An effective weapon to fight bacterial pathogens is the naturally present lytic enzymes, which selectively target particular structures of bacterial cell wall and cause rapid cell lysis likely without triggering common resistance. Lytic enzymes, usually consisting of a catalytic domain and a cell wall binding domain (CBD), can be engineered by domain swapping for the generation of new functions. In addition, these enzymes, when immobilized onto or incorporated into nanomaterials, retain excellent bactericidal activity and stability either in solution phase or on solid surfaces. These properties make lytic enzymes promising candidates in supplementing or replacing traditional antibiotics for environmental disinfection/decontamination or as therapeutics. In this talk, three specific topics will be covered. First, various enzyme-nanoconjugates with excellent antibacterial activities against several common food-borne or hospital-acquired bacterial pathogens will be showcased, with a focus on silver nanoparticle-CBD conjugates, which selectively kill *Bacillus anthracis* in a mixture with *B. subtilis* or *Staphylococcus aureus*, while silver nanoparticles alone are less active and non-selective. In the second part, engineering of lytic enzymes for sensitive detection of a single bacterial species in a multi-species mixture will be described. Biotin-CBDs, streptavidin (SA), and biotinylated reporters are self-assembled into CBD-SA-reporter complexes, which enable simultaneous spectrophotometric detection of *S. aureus*, *B. anthracis*, and *Listeria innocua* cells in salt solution with a detection limit of >100 CFU/mL. Further modification of this system results in self-assembled CBD-SA-DNA complexes, which achieve a 1-10 CFU/mL detection limit without cross-reactivity when combined with qPCR. In the third part of this talk, the basic understanding of the functionalities of lytic enzymes in complex, non-ideal environment will be discussed. The lytic enzyme CD11, being highly active in buffer but inactive in rich growth media, efficiently kills *Clostridium difficile* with a decreased level of the wall teichoic acid (WTA), a bacterial cell surface glycopolymer. The low activity of lysostaphin against *S. aureus* in unprotonated amines is also regulated by the WTA. A possible mechanism of WTA-mediated cell resistance is proposed.

530039 Nanobiocatalytic Antifouling in Wastewater Treatment Via Quorum Quenching

Abstract:

Enzymes can be employed as an effective tool to solve a variety of environmental problems such as membrane biofouling, microbial contamination, and even global warming due to CO_2 emission. However, the poor stability of enzymes often hampers the practical uses of enzymes for a long-term operation. Nanobiocatalysis, using nanostructured materials for the enzyme immobilization, has gathered a growing attention due to its unprecedented successes in improving the enzyme stability as well as enzyme loading. This presentation will introduce the recent developments of nanobiocatalysis with a quorum-quenching acylase for membrane antifouling in wastewater treatment. Acylase catalyzes the hydrolysis of signal molecules for quorum sensing among microorganisms, which plays an important role in the formation of microbial biofilms. Acylase was immobilized and stabilized in/on nanostructured materials via nanobiocatalytic approaches for their successful antifouling of membrane via quorum quenching mechanism. Several representative examples of nanobiocatalysis for effective antifouling will be presented in detail.

533155 Nano-Bio-Catalysts for Enzymatic Biofuel Cells

Abstract:

Carbon based materials with nanostructures as a support for enzyme immobilization hold a great potential to fabricate the enzyme electrode that meet both high power density and long-term stability requirements for the successful applications of enzymatic biofuel cells. In this paper, we will immobilize our model anode and cathode enzymes over various carbon nano-materials to form high performing enzyme electrodes via nano-confinement and/or multipoint crosslinking approaches. In terms of achieving the high power density output, we will explore a large surface area of carbon nano-materials for improving the enzyme loading per a unit geometrical surface area of electrodes for enzymatic biofuel cells. By increasing the enzyme loading, the available active sites for electron generation also increases in case of the enzyme anode. As the more number of electrons are generated, a probability of â€œcollectingâ€ them at the anode current collector would increase as well. Furthermore, carbon based nano-materials possess a high electronic conductivity, which can enhance the electron transfer rate of enzyme electrodes by decreasing its ohmic overpotential. In terms of achieving the high stability, we will explore the nano-confinement and/or multipoint crosslinking effect of enzymes. The electrochemical activities and long-term stability of these enzyme electrodes will be evaluated and correlated to their enzyme immobilization process parameters and their structures. Based on this performance-structure relationship, we will further optimize their electrochemical performances toward the biofuel cell applications.

537827 Surface Active Biological Agents for Fabrication of Functional Materials for Biomedical Applications

Abstract:

Functional materials are highly desired for a wide range of applications. Biological agents are particulary appealing since they offer high specificity and a great variety of functionality. The presentation reviews our recent efforts in designing and producing functional biological agents for fabrication of materials. We showed that pH and enzyme acivity-responding coatings on porous nanoparticles could afford efficient controlled drug release, while single chain antibodies could enable cell recognition on drug-loaded nanoparticles. We further showed that materials such as plastics and cotton fabrics can be efficiently modified by application of hydrophobin proteins, or cellulose binding domains

530077 Engineering Nanoscale Protein Scaffolds with Modular Functionalities

Abstract:

Proteins are the most versatile among the various biological building blocks and efforts in protein engineering have resulted in many industrial and biomedical applications. However, the strength of proteins - their versatility and specific interactions - also complicates and hinders their systematic design and engineering. One potential approach is to exploit the modular nature of different protein domains in order to design synthetic protein scaffolds that can perform completely new biological functions. In this talk, I will outline several successful examples in connecting exchangeable protein domains for predicative engineering applications in (1) bioenergy applications, (2) Fuel cell applications, and (3) biosensing.

537002 Characterizing Micellar Assembly of Oligonucleotides with Polyelectrolytes

Abstract:

Polyelectrolytes serve as robust building blocks for self-assembled nanoparticles and customizable materials. Neutral-charged block copolymers can assemble with charged macromolecules to form micelles with a dilute neutral block corona. Proteins encapsulated with such polymers are protected and stabilized and can be released under programmed conditions. Still, there is much to learn about the material properties and dynamics of these hybrid assemblies and how they react to mechanical and chemical signals. This work uses biocompatible copolymers to encapsulate charged proteins into micelles with programmable functionality. A library of polymers and proteins allows us to define density and complexity of our assemblies as we study stability and tune the release of cargo under a range of conditions. Investigating physical properties and kinetics will help develop a system useful for triggered release of therapeutic payloads.

537360 The Binary Effect on Drug-Resistant Bacteria of Polymeric Vesicles Appended By Proline-Rich Amino Acid Sequences and Inorganic Nanoparticles

Abstract:

Prevalent research underscores efforts to engineer highly sophisticated nano-vesicles that are functionalized to combat antibiotic-resistant bacterial infections, especially those caused by methicillin-resistant *Staphylococcus aureus* (MRSA), and that aid with wound healing or immunomodulation. This is especially relevant for patients who are susceptible to *S. aureus* infections post-operatively. Technical challenges associated with this aim require a thorough assessment of the chemical properties of synthesis materials and methods. Here, formulations were incorporated into polymeric, biocompatible vesicles called polymersomes that self-assemble via hydrophobicity interactions of admixed aqueous and organic substances. Nano-polymersomes were synthesized using a high molecular weight amphiphilic block copolymer, and were conjugated to include antimicrobial peptides (AMPs) along the peripheral hydrophilic region and silver (Ag) nanoparticles inside their hydrophobic corona. *In vitro* testing on bacterial and human cell lines indicated that finely tuned treatment concentrations of AMP and Ag nanoparticles in polymersomes synergistically inhibited the growth of MRSA without posing significant side effects, as compared with other potent treatment strategies. In particular, a ratio of silver-to-AMP of about 1:1.9, corresponding to approximately 0.1 mg/ml of silver nanoparticles and 40 Î¼M of the peptide, yielded complete MRSA inhibition over a 23-hour time frame. The silver nanoparticles generate bacterial cell wall indentations that prompt the influx of external entities and the outflow of vital proteins or nutrients. This weakening of lipopolysaccharide permeability barriers promotes the unimpeded cellular translocation of contiguous AMP molecules, which can then disrupt internal processes including DNA or protein synthesis. This bacteriostatic activity, coupled with nominal cytotoxicity towards native human dermal fibroblast cells, extends the potential for AMP/Ag nanoparticle polymersome therapies to replace antibiotics in the clinical setting. This study will facilitate contemporary research that is feasible, important, and highly relevant, and possesses the potential to advance knowledge in several fields, including drug delivery, nanotechnology, and artificial immunity.

535010 Generation of Plasmonic Nanoparticles in an Amino Acid Incorporated Hydrogel for Detection of Low Doses of Ionizing Radiation

Abstract:

Radiotherapy has shown immense therapeutic value to treat cancerous lesions inside the body. The radiation treatment of a patient is a multi-stage complex process that requires a high level of accuracy. Even with the quality assurance that is performed prior to treatment, there are significant causes for failure in delivering the prescribed radiation dose including equipment failure and operating error. The existing technology to measure these doses requires trained personnel to operate and complex read out mechanisms making their routine use sparse. To alleviate these concerns we have developed a colorimetric sensor that changes color upon exposure to ionizing radiation. We use the free radical species generated through radiolysis of water to reduce the metal ions to metal atoms which further nucleate and grow into nanoparticles templated by amino acids. This sensor has been incorporated in a hydrogel to mitigate handling and storage concerns. The generated nanoparticles in the hydrogel were quantified through Ultraviolet-Visible Spectroscopy. We observed that the yield of nanoparticles was proportional to the radiation dose and was used to quantify the delivered radiation dose. Transmission Electron Microscopy and Dynamic light scattering was used to quantify the size of the nanoparticles. Finally, we show the sensors capability in a canine patient that was undergoing treatment. Our predictions using our sensor matched the expected prescribed dose. The ease of operation of the colorimetric sensor makes it a potential candidate during clinical radiotherapy.

530693 Unusual Electronic Properties of Template-Directed π-Conjugated Porphyrin and Phosphorene Nanotubes

Abstract:

Using large-scale DFT calculations (up to 1476 atoms and 18,432 orbitals), we present the first detailed analysis on the unusual electronic properties of recently synthesized porphyrin[1] and phosphorene nanotubes.[2] We surprisingly observe extremely large oscillations in the bandgap of these nanostructures as a function of size, in contradiction to typical quantum confinement effects (i.e., the bandgap increases with size in several of these nanotubes). In particular, we find that these intriguing electronic oscillations arise from a size-dependent alternation of aromatic/nonaromatic characteristics in these porphyrin nanotubes. Our analyses of band structures and orbital diagrams indicate that the electronic transitions in these nanostructures are direct-bandgap, optically active bright states that can be readily observed in photoelectron spectroscopic experiments. Most importantly due to their unusual bandgap oscillations, we find that both type I and type II donor-acceptor p-n heterojunctions are possible in these template-directed, bottom-up synthesized, porphyrin nanotubes - a unique property that is not present in conventional carbon nanotubes.

[1]Sarah I. Allec, Niranjan V. Ilawe, and Bryan M. Wong
"Unusual Bandgap Oscillations in Template-Directed π-Conjugated Porphyrin Nanotubes."
Journal of Physical Chemistry Letters, 7, 2362 (2016)

[2]Sarah I. Allec and Bryan M. Wong
"Inconsistencies in the Electronic Properties of Phosphorene Nanotubes: New Insights from Large-Scale DFT Calculations."
Journal of Physical Chemistry Letters, 7, 4340 (2016)

538032 Modeling the Aggregation Behavior of Cyanine Dyes for Efficient Energy Transport

Abstract:

In natural photosynthetic systems, quantum coherent effects contribute to enhanced efficiency in energy transfer pathways when light-harvesting chromophores are closely packed or aggregated. We can mimic these effects synthetically by exploiting the J- and H-aggregation behavior of certain dyes, such as cyanines. A proof-of-concept study has established that J-aggregated pseudoisocyanine (PIC) dyes can be rationally attached to DNA scaffolding, where they self-assemble in the minor groove of continuous A-tract DNA sequences. In addition, in concentrated aqueous solution PIC forms homogeneous fibers with an average width of 2.89 nm, shown from a cryo-TEM analysis. To determine the physical mode of aggregation in aqueous solution, we use density functional theory (DFT) to analyze the excitonic transport properties of the model PIC aggregates to understand how the physical aggregation behavior of PIC influences its energy transfer efficiency. We note that a chiral-aggregate model in which PIC monomers are neither parallel nor orthogonal relative to the long-axis of the fiber replicates the experimental spectra most accurately, where the J-aggregate absorption band is red-shifted ~1600 cm^{-1} from the monomer band. This model of aggregation could be formed in concentrated aqueous solution by the sequential binding of PIC dimers and monomers to the ends of the fiber. Also, H-aggregates of cyanine dimers and trimers can be engineered using covalent attachment to the backbone of DNA nanostructures. Using molecular dynamics, the H-aggregated structure of cyanine dimers and trimers can be analyzed, leading to improved estimates of spectral and excitonic properties of these systems. Both J- and H-aggregated templated dye molecules studied here can be further utilized in the rational design and optimization of DNA-dye excitonic circuits using PIC and related cyanine dyes.

536287 Spatial Tailoring of Dopant Position in Solids for Enhanced Visible Light Photocatalytic Performance

Abstract:

Transition metal (TM) doped semiconductor materials are extensively employed for light harvesting and photocatalytic applications to increase the charge mobility for better performance. The position and site occupancy of the dopant dictate the physical properties of these materials such as light absorption, separation of charge carriers, and the surface reaction kinetics. Commonly used bulk characterization techniques are often indifferent to surface vs bulk dopant occupancy. Therefore, this work addresses the phenomenon of spatially controlled dopant incorporation for controlled optical and photocatalytic properties.

In this work, highly doped TiO_2:Ni^{2+} (15 mol%) nanoparticles are synthesized via sol-gel chemistry. The drying and annealing of the aged sol are shown to affect the segregation of NiO onto the surface of TiO_2. Specifically, it is possible to control the dopant position by varying the moisture exposure time, indicating that the ambient water layer on the surface of TiO_2 plays an important role on surface energetics. The effect of moisture and annealing rate on the dopant position was systematically studied using steady state and time resolved methods (XRD, UV-Vis, Raman, TGA/DSC, and FTIR) to extract crystallization temperatures, optical absorption, bond formation, and surface hydroxyl concentration. XRD results show that the vacuum dried TiO_2:Ni powders formed a doped anatase phase while the air dried powders formed segregated anatase and NiO phases upon annealing. Furthermore, rapid annealing of the air dried TiO_2:Ni powders formed a metastable anatase doped phase while slow annealing resulted in the segregation of NiO phase, which can also be observed in TGA/DSC. The annealing processes result in a shift from discrete peaks to a broad absorption based on the dopant position. The evolving local environment is probed via x-ray absorption spectroscopy (XAS) and High-Resolution TEM to elucidate the difference in the dopant local environment with photocatalytic performance using photoluminescence (PL) spectrometer. Moreover, the charge carrier recombination kinetics in these NPs can investigated using time-resolved PL measurements to understand the reaction pathways. This ability to incorporate higher dopant concentrations while engineering the dopant position in nanostructures will assist in the development of improved solar cells and photocatalytic devices. Finally, a similar trend of dopant segregation was observed with other first row TM doped TiO_2 powders upon slow annealing, suggesting that the segregation of TM dopants in TiO_2 is a function of the dopant size.

536577 Enhancement of Photocatalytic Reduction Reaction on TiO_2 Under Solar Light Using Alternative Plasmonic Titanium Nitride Nanoparticles

Abstract:

In this work, we report plasmon enhanced photocatalytic reduction reaction on TiO_2 under solar light using alternative plasmonic titanium nitride nanoparticles. The excellent chemical stability at elevated temperatures and higher integrated absorption efficiency under the solar light in comparison to conventional plasmonic nanomaterials (silver and gold) make titanium nitride a promising candidate for the application. We observed an efficient production of formic acid through simultaneous photoreduction of bicarbonate and oxidation of glycerol in the presence of TiO_2/TiN composite nanocatalysts in an isothermal condition. The photocatalytic productivity of formate was 465 mM /g cat-hr using TiO_2/TiN composite, which is significantly (15 times) higher than TiO_2 alone. Most importantly, the productivity of formic acid using TiN/TiO_2 nanoparticle catalysts was found to be higher than that of similar Au/TiO_2 and Ag/TiO_2 composites. The enhancement is predicted to be due to the plasmon-induced excited charge transfer from TiN to the conduction band of TiO_2. We also found that TiN can efficiently convert excited plasmons into heat. In an adiabatic condition under solar illumination, the temperature of reactant solution can increase as much as up to 20°C from room temperature. Interestingly, at that elevated temperature, the enhancement effect of TiN is negligible. Presently comprehensive theoretical analyses in concert with transient absorption spectroscopy are in progress to unravel the exact mechanism behind our observation.

527491 ZnO Nano Forest As Electrode Material for DSSC: Where Is the Bottleneck?

Abstract:

The efficiency of dye sensitised solar cells has not improved substantially over the last decade. To resolve this problem, different electrode materials are being tried. One of the promising electrode material is ZnO which has orders of magnitude higher electron mobility than TiO_2. Another advantage of ZnO is that it can produce a variety of morphologies such as nano-rods, wires, forests, flower, tetrapods easily. Such structures, especially the nano-forest, reduces the chances of electron scattering to minimum by providing a network of electron paths of gradually changing hierarchy.

Although synthesis of ZnO nano forest structure has been demonstrated by hydrothermal process, the growth rate is painstakingly slow for this process and may not be suitable for practical purpose. In this talk we shall discuss a microwave based process for quick production of the nano forest and hence resolving this production bottleneck. The new strategy of force seeding for quick and efficient branch growth will also be discussed.

Even when electron freeways and avenues practically eliminates electron scattering at the particle grain boundaries, the efficiency of ZnO solar cells are much lower than those of TiO_2. In this work, we will show that the reason for this bottleneck is limited dye adsorption. The most prevalent Ruthenium based dye N719 is acidic enough to corrode the electrode surface and hence result in poor dye uptake. Microscopic investigation of the electrode surface reveal a much severe corrosion to the nano-forest structure due to small size of the branches.

It will be shown that soaking the electrode in a very concentrated dye solution for very short period may lead to some adsorption. However, huge dye loss occurs in subsequent washing steps. Nevertheless, within these constraints, the dye adsorption has been optimised to keep the efficiency of the cell to reasonable value. EDX and microscopy reveals that a very small fraction of the surface is actually covered with dye and hence dyes specifically suitable for ZnO should be used for better efficiency. Results for less acidic dyes will also be discussed.

534139 Ultrathin Plasmonic Coatings for Selective Radiative Transmission in Silica Aerogels

Abstract:

Current solar thermal collectors use vacuum insulation which only partially blocks outgoing infrared radiation from the absorbing surface. The losses associated with this phenomenon lead to a decrease in collector performance due to a reduction in temperature or efficiency. Visibly-transparent silica aerogels are a promising alternative to vacuum insulation in solar thermal applications. However, they are also characterized by undesirable transmission in the mid-infrared band. Here, we investigate ultrathin conformal plasmonic coatings to selectively enhance the absorption of silica aerogels in this target band and suppress outgoing radiative losses at high temperatures. We fabricated aerogels coated with aluminum-doped zinc oxide (AZO) using atomic-layer deposition (ALD). An evaluation of the microstructure via electron microscopy confirms that ALD is a suitable technique for the conformal coating of aerogels. Furthermore, we characterize the effect of the AZO coatings on the optical and thermal properties. We demonstrate AZO-coated aerogels with high solar transmission and low infrared transmission. The absorption coefficient in the target mid-infrared band increases substantially due to the AZO. In contrast, transmission of solar radiation through the materials was not significantly impacted. Ultrathin plasmonic coatings have the potential to significantly improve the performance of low-irradiance, high-temperature solar collectors.

523945 Computational Electrochemistry of DNA: Effect of Lithium

Abstract:

The general electrochemistry of deoxyribose nucleic acid (DNA) is studied using first-principles computational methods. For developing a systematic approach for analysis of its electrochemical capabilities, DNA is divided into three structural units of interest: the nucleobase, the deoxyribose sugar, and the phosphate group. Through this parted approach, in studying nucleobases, nucleosides, and nucleotides as separate species, it was possible to elucidate each contributive effect. As one of the key insights into electrochemistry, reduction potentials were pursued, via density functional theory (DFT) modeling with PBE0/6-31G**+ calculations. Furthermore, given that solvents are often involved in electrochemical applications, the effect of solvation on the reduction potential of DNA is characterized using implicit solvent model. From this study, we find that the reduction potential is affected by the solvation through conformational change. From this study, new insights are offered on the relationship between chemical structures and redox properties in DNA, promoting fundamental understanding of the electrochemistry for biological molecular materials.

539431 Facile and Scalable Fabrication of Sulfur Cathodes Via Air-Controlled Electrospray

Abstract:

Lithium-sulfur batteries (Liâ€"S) are rising as one of the next-generation energy storage devices because of their high theoretical capacity and energy density. In addition, this battery system utilizes sulfur as the active material, which is non-toxic and abundant. Despite these striking advantages, there are some limitations to this technology that remains unsolved. Herein, we propose a facile and effective strategy to construct three-dimensional porous electrodes for Li-S battery applications. Through the means of air-controlled electrospraying process, cathode materials can be coated uniformly onto the current collector in the absence of any cracking or pinholes, and the thickness and morphology of electrodes were able to be controlled easily by tuning the experimental condition such as deposition time, electrical field, feeding rate, solution concentration *etc*. Electrochemical evaluations have revealed that the improved cycling and rate performance were attributed to the 3D open porous structure with well-developed conductive network, suggesting high prospects of this coating approach being applied in Li-S batteries.

523676 Direct Conversion of CO_2 to Carbon Materials for Energy Conversion and Storage

Abstract:

Porous carbon material can be synthesized from sodium borohydride ($NaBH_4$) and carbon dioxide (CO_2) under an atmospheric pressure and moderate temperature. The synthesized carbon material can be used as an advanced electrochemical energy material, such as active electrocatalysts for oxygen reduction reaction (ORR) and electrode materials for supercapacitor or secondary battery. The composites of the carbon material with metal oxides exhibited enhanced electrochemical properties when applied in various electrodes due to increased specific surface area, improved electrical conductivity, and stable interaction between carbon material and metal oxides.

527051 Mn_3O_4Â Nanoarray and Solid Electrolyte Interface Encapsulated Nanoarray Electrodes for High Performance Lithium Sulfur Battery

Abstract:

Mn_3O_4 nanoarray and solid electrolyte interface encapsulated nanoarray electrodes for high performance lithium sulfur battery

Junling Guo, Xiaolong Zhang, Xinyu Du, He Yang and Fengxiang Zhang

School of Petroleum and Chemical Engineering, Dalian University of Technology

2 Dagong Road, Liaodongwan New District, Panjin, China, 124221

Lithium-sulfur batteries (LSBs) are promising next generation technology to tackle the worldwide energy and environmental issues owing to their high theoretical energy density (2,500 $kWkg^{-1}$), low cost and environmental friendliness of sulfur as the active species at cathode. However, practical application of LSBs is restricted by its poor cycle life, which stems mainly from the "redox shuttle reactions" of dissolved polysulfides. Herein, we report a novel cathode structure based on nano-wall-array Mn_3O_4 which shows excellent cycle- and rate performance. The Mn_3O_4 nano-wall arrays function as "nano reservoirs" for sulfur confinement so that the cathode can yield a high-rate (2C) initial- (~593 mAh/g) and reversible capacity (~355 mAh/g, 60 % retention after 3000 cycles).

To provide another possibility of suppressing polysulfide shuttle and further improve the cycle performance of LSB, we explore encapsulation of sulfur and electrolyte together in CNT array by the solid electrolyte interface (SEI), which could selectively sieve Li^+ ions while efficiently avoiding polysulfide accumulation and suppressing undesired polysulfides migration. Our strategy is simple, straightforward and effective. It just needs cycling the carbon/sulfur cathode a few times within a potential window of 0.3-1.0 V to form such a smart SEI. The LSB assembled with such a cathode shows very good rate perforamene (1275, 1081, 889 and 667 mAh g^{-1} at 0.2, 0.5, 1 and 2C, respectively), high Coulombic efficiency (~99 % after 4 cycles) and stable cycle performance (81.4 % capacity retention after 200 cycles for SEI wrapped electrode and 54.5% after 50 cycles for pristine cathode.

528080 Gyroidal 3-D Electrochemical Energy Storage Nanoarchitecture

Abstract:

Gyroidal mesoporous carbon monoliths are employed to integrate all electrochemical energy storage device components, namely anode, electrolyte, and cathode, into a single three-dimensionally (3â€'D) interdigitated nanoarchitecture. Integrating electrodes and electrolyte into a 3-D interdigitated network structure allows for small ion diffusion distances, while allowing phase connectivity and preventing wasteful porosity. Hence, such 3-D electrical energy storage devices have the potential for structure-driven power enhancements, but have remained challenging to achieve on the nanoscale. The lack of methods for fabricating and assembling the appropriate functional materials in 3-D co-continuous nanoarchitectures is a key issue holding back the development of such intricate 3â€'D device architectures. The components in electrochemical energy storage systems have stringent requirements on their functionality; the electrodes need to display electrical conductivity in combination with redox-activity with an electrochemical potential difference between anode and cathode, and the electrolyte needs to be ionically conductive but electronically insulating. In 3-D interdigitated designs, the individual components cannot be fabricated separately and subsequently assembled or layered, but synthesis and assembly are inherently connected. In particular, the nanoconfined synthesis of functional materials in the presence and close proximity of the other functional device components necessitates synthetic procedures that are compatible between all materials. Here we present pathways to such electrochemical energy storage systems with all functional phases integrated in a triblock terpolymer derived core-shell double gyroid morphology. We devised a step-wise bottom-up strategy to synthesize and incorporate all functional materials within each other in a compatible way. Our 3-D interdigitated design comprises a gyroidal carbon anode network that is electrochemically coated with an ultrathin polymer electrolyte in a self-limiting and -regulating method. The 3-D continuously nanoporous anode-electrolyte monolith is in turn filled with a sulfur-polymer hybrid cathode that is polymerized in the confinement of the nanoporous network and exhibits lithium redox-activity and electrical conductivity. All phases are less than 20 nm in their layer dimensions and integrated throughout a macroscopic monolith. Despite the small separation distances, the solid-state nano-3-D energy storage system exhibits battery-like characteristics and can be cycled numerous times.

530798 Modeling Mechanisms of Nickel Oxide Lithiation Using First Principles Calculations and Classical Nucleation Theory

Abstract:

Metal oxide conversion reactions present an opportunity for ultrahigh high capacity lithium ion battery electrodes, wherein the direct reaction of lithium ions with lattice oxygen anions leads to a discharge product consisting of metal nanoparticles embedded in a matrix of amorphous lithia (a-Li_2O). Despite the promise of such chemistries, there exist many challenges toward their practical implementation. Of primary concern for these materials is the presence of large overpotentials, which lead to significant voltage hysteresis between charge and discharge cycles.[1] For example, nickel oxide (NiO) lihiates at 0.6 V, whereas the bulk reduction potential is at 1.86 V.[2] To address these concerns, nanoscale interfacial engineering has been shown to enhance the reversibility of conversion reactions and reduce overpotentials.[3,4] In this work, we use density functional theory (DFT) and classical molecular dynamics (MD) calculations to understand interfacial reactions for the NiO conversion reaction. DFT-calculated interfacial energies for Ni|a-Li_2O, Ni|NiO, and NiO|a-Li_2O interfaces are performed, and integrated into a classical nucleation theory (CNT) model. This DFT-MD-CNT computational approach is used to calculate potential-dependent nucleation barriers and is compared to structural details obtained from *in operando* synchrotron X-ray scattering experiments on multilayer Ni|NiO thin film electrodes. In agreement with experimentally observed high voltage changes in electron density at the multilayer interface, we determine that a lithium-rich region at the Ni|Li_2O interface significantly lowers the predicted nucleation barrier. This result is consistent with mechanisms for capacitive interfacial charge storage in other conversion reaction materials.[5]

This research was supported as part of the Center for Electrochemical Energy Science, an Energy Frontier Research Center funded by the U.S. Department of Energy, Office of Science, Office of Basic Energy Sciences, under Contract No. DE-AC02-06CH11357.

References

1. J. Cabana, L. Monconduit, D. Larcher, and M. R. Palacín, *Adv. Mater.*, **22**, E170–E192 (2010).
2. P. Poizot, S. Laruelle, S. Grugeon, L. Dupont, and J.-M. Tarascon, *Nature*, **407**, 496–499 (2000).
3. G. Evmenenko et al., *ACS Appl. Mater. Interfaces*, **8**, 19979–19986 (2016).
4. G. Evmenenko et al., *Phys. Chem. Chem. Phys.*, **19**, 20029–20039 (2017).
5. Y.-Y. Hu et al., *Nat. Mater.*, **12**, 1130–1136 (2013).

539387 Development of a New Generation of Stable, Tunable, and Catalytically Active Nanoparticles Produced By the in-Situ and Ex-Situ Synthesis Methods

Abstract:

Nanoparticles (NPs) are revolutionizing many areas of science and technology, often delivering unprecedented improvements to properties of the conventional materials. However, despite important advances in NPs synthesis and applications, numerous challenges still remain. Development of alternative synthetic method capable of producing very uniform, extremely clean and very stable NPs is urgently needed. If successful, such method can potentially transform several areas of nanoscience, including environmental and energy related catalysis. Here we present the latest developments in demonstration of catalytically active NPs synthesis achieved by the helium nanodroplet isolation method. This alternative method of NPs fabrication and deposition produces narrowly distributed, clean, and remarkably stable NPs. The fabrication is achieved inside ultralow temperature, superfluid helium nanodroplets, which can be subsequently deposited onto any substrate. This technique is universal enough to be applied to nearly any element, while achieving high deposition rates for single element as well as composite coreâ€"shell NPs. In addition, our work also involves development of a new generation of doping-segregation method on nanoparticle synthesis. Here we studied behavior of dopants by various in-situ characterization techniques including environmental TEM, complemented by such in-situ techniques as XAFS, XRD and DRIFTS. More specifically, we investigated the effect of such conditions as reduction temperature on the size of nanoparticles. We then applied this method to prepare very active and selective catalysts for CO_2 reduction by reverse water gas shifts reaction.

535230 Switchable Surfactants for the Preparation of Monodisperse, Supported Nanoparticles and the Effects of Calcination on Nanoparticle Characteristics

Abstract:

Traditional synthetic methods of preparing supported nanoparticle catalysts require the use of stabilizing ligands that passivate the nanoparticle surface to maintain size and monodispersity. Stabilizing ligands bound to the nanoparticle surface compete with reagents for active sites and typically significantly hinder catalytic activity. Commonly used methods of ligand removal, such as calcination, have detrimental effects on the catalyst as exposure to high temperatures often results in significantly increased nanoparticle size and, thus, decreased total surface area for catalysis. Previously, we demonstrated a novel method for synthesizing highly active, monodisperse, supported nanoparticles using a switchable surfactant (SwiS) system. In this method nanoparticle size is finely controlled throughout synthesis and deposition. Here, we show with X-ray photoelectron spectroscopy (XPS) that supported nanoparticles prepared with SwiS are completely surface-clean after deposition, eliminating the need for any traditional activation steps such as calcination. Additionally, it is demonstrated that a low-temperature calcination at 230°C of surface-clean supported nanoparticles has detrimental effects on nanoparticle size, dispersion, and catalytic activity. Supported nanoparticles prepared with SwiS are up to 700% more active in the hydrogenation of 4-nitrophenol than their calcined counterparts. Further, calcination results in the formation of an induction time in 4-nitrophenol reduction, demonstrating that calcination causes surface structure rearrangements unfavorable to catalysis.

530819 A Commercially-Viable One-Step Synthesis Method to Prepare MWW Zeolite Nanosheets

Abstract:

Zeolites are shape-selective microporous crystals that have been widely used as heterogeneous catalysts in petroleum and chemical industries.[1] However, pore dimensions (4-7 Å) of typical zeolites often encounter diffusion limitations, especially in applications that involve bulky molecules. MCM-22 (MWW type), one of the most investigated two-dimensional zeolites, exhibits properties that can alleviate mass transport-related issues by providing strong accessible acid sites on its exterior surface.[2] Conventional layered MCM-22 precursors, or MCM-22(P), result in the direct condensation of MWW layers during calcination, sacrificing significant accessibility to surface acid sites. In this talk, we will discuss an alternative method to avoid layer condensation.

The ability to generate layered MCM-22 traditionally requires post-synthesis treatment with an exfoliating agent, such as surfactants, to disorientate layers of MCM-22(P).[3] In this talk, we will describe how we have modified the synthesis of MCM-22 to directly yield MWW nanosheets via a single step, as determined by a combination of powder X-ray diffraction (XRD), transmission electron microscopy (TEM), and gas adsorption measurements. This modified synthesis method increases the external surface area to 257 m^2/g as compared to conventional MCM-22 layered materials (106 m^2/g). In addition, the coordination environment of Al isotopes was investigated by ^{27}Al magic-angle spinning nuclear magnetic resonance (MAS NMR) spectroscopy after calcination, revealing that the amount of extra-framework Al species in disordered MWW is 7 mol% less than the ones of ordered MWW, thus suggesting the optimized one-step method helps to prevent dealumination. To confirm the advantages of creating higher surface area nanosheets, we performed a cumene cracking reaction to evaluate the catalytic activity of the delaminated MWW, which has a longer lifetime than the ordered one. Collectively, our studies reveal an innovative way to prepare delaminated MWW catalysts for commercial applications in the (petro)chemical industries.

1. Davis, M. E.; *Nature.* 417 **(2002)** 813-821

2. Roth, W. J.; Nachtigall, P.; Morris, R. E.; ÄŒejka, J.; *Chem. Rev.* 114 **(2014)** 4807-4837

3. Xu, L. et al.; *Chem. Mater.* 28 **(2016)** 4512-4521

534238 Photocatalytic Inorganic Core Hedgehog Particles

Abstract:

In order to disperse inorganic catalytic nanoparticles in apolar media, chemical camouflage is required for dispersion stability, which greatly reduces catalytic activity. Due to this barrier, there has been little investigation of inorganic catalysts in apolar solvents. This study seeks to create functional catalysts utilizing surface corrugation to circumvent this requirement. Hedgehog Particles (HPs) recently developed in the Kotov lab are polystyrene microparticles with nanoscale zinc oxide spikes which result in marked reduction of Van der Waals interactions and enable dispersion in both polar and nonpolar solvents. HPs have been expanded to inert and active inorganic core materials including silica and hematite by utilizing layer-by-layer polyelectrolyte films. The inorganic core HPs maintain dispersion in a wide array of nonpolar environments and exhibit superior chemical stability and thermal stability. Inorganic HPs display enhanced photocatalytic activity compared to nanorods in the degradation of methylene blue in water. Additionally, photocatalytic activity in chloroform is achieved by the templating of zinc oxide spikes to form titanium dioxide spikes indicating promise for use in nonpolar processes such as cyclohexane oxidation. Inorganic core HPs represent a new family of dispersible catalysts in nonpolar environments.

520663 Slowing the Kinetics of Alumina Sol-Gel Chemistry for Controlled Catalyst Overcoating and Improved Catalyst Stability and Selectivity

Abstract:

Catalyst overcoating is an emerging approach to engineer controlled surface functionalities on supported metal catalyst and improve catalyst selectivity and durability. Alumina deposition on high surface area material by sol-gel chemistry has traditionally been difficult to control due to the fast hydrolysis kinetics of aluminum-alkoxide precursors. Here, we adapt sol-gel chemistry methods to slow down these kinetics. Specifically, we report two novel liquid phase sol-gel methods to deposit nanometer-scale alumina overlayer with similar chemical properties but distinct physical textures, onto high surface area substrates. The alumina overcoats are comparable in conformality to the overcoats prepared from atomic layer deposition. Our strategy relies on regulating the hydrolysis/condensation kinetics of $Al(^{s}BuO)_3$ by adding a chelating agent or using non-hydrolytic sol-gel chemistry. With chelation chemistry, a conformal yet porous overcoat leads to a highly sintering-resistant Cu catalyst for liquid-phase furfural hydrogenation. With the non-hydrolytic sol-gel route, a denser Al_2O_3 overcoat can be deposited to create a high density of Lewis acid-metal interface sites over Pt/SBA-15. This overcoated Pt/SBA-15 has a substantially increased hydrodeoxygenation activity for the conversion of lignin-derived 4-propylguaiacol into propylcyclohexane with up to 87% selectivity.

526287 One Step, Steady State Catalytic Conversion of Methane to Methanol Using Copper Zeolites: Kinetics and Site Requirements

Abstract:

Abundant stores of natural gas have driven desire for effective utilization of CH_4 within the current petrochemical infrastructure. Methane-derived methanol production typically proceeds through a high temperature (~1173 K), energy-intensive two step upgrading process. This work demonstrates a low temperature (~493 â€“ 573 K), single step, steady state methanol production process over copper-based zeolitic catalysts. We utilize transient and steady state kinetic measurements as well as in situ titration in conjunction with infrared spectroscopic experiments in order to propose a C-H activation mechanism, subsequent conversion pathways, and reaction energetics as well as providing methods to identify and quantify active site motifs. The current work represents the first kinetic investigation and characterization of Cu-based catalysts for the steady state conversion of methane to methanol in zeolitic catalysts.

538498 Dry Reforming of Methane over $Ce_{0.7}Ti_{0.3}O_{2-\Delta}$ supported Nickel Catalyst

Abstract:

Dry reforming of methane over $Ce_{0.7}Ti_{0.3}O_{2-\delta}$ supported nickel catalyst

Sachin U. Nandanwar,[a,b] Yunkai Zou,[a] Linze Du,[a] Joseph H. Holles,[b] Jing Zhou,[a]

[a]*Department of Chemistry, University of Wyoming, Laramie, WY 82071, and Email address: jzhou2@uwyo.edu*

[b]*Department of Chemical Engineering, University of Wyoming, Laramie, WY 82071, and Email address: jholles@uwyo.edu*

Abstract

Dry reforming of methane (DRM) utilizes two abundantly available greenhouse gases to produce industrially important syngas ($CH_4 + CO_2 \rightarrow 2CO + 2H_2$). Syngas is used to produce synthetic petroleum as fuels or chemicals. The conversion of methane, the primary component of natural gas, to higher value products will become increasingly important for the near future. Current global challenge for the DRM reaction is to develop thermally stable and active catalysts that can show good resistance to deactivation. In our study, $Ce_{1-x}Ti_xO_{2-\delta}$ (x: 0.1, 0.2, 0.3, and 0.5) supported nickel catalysts were synthesized by sol-gel methods with controlled Ni loadings between 1 and 10 wt.%. Ceria exhibits unique redox properties and oxygen storage capacities, which can better anchor Ni as small clusters and inhibit coke formation. Introduction of metal dopant, Ti, into ceria could promote its redox properties as well as enhance its thermal stability at high temperatures. All the prepared catalysts were characterized by a combination of spectroscopy and microscopy techniques including X-ray diffraction (XRD), Transmission Electron Microscopy (TEM), Scanning Electron Microscopy-Energy Dispersive Spectroscopy (SEM-EDS), X-ray Photoelectron Spectroscopy (XPS), BET surface area, H_2-chemisorption, and H_2-temperature-programmed reduction (H_2-TPR). Their DRM performance was investigated in a continuous gas flow reactor using gas chromatography and mass spectrometry instruments and compared with respect to reaction temperatures, Ce:Ti ratios in the oxide supports, and Ni loadings. XPS and XRD data show the formation of $Ce_{1-x}Ti_xO_{2-\delta}$ solid mixed oxide supports. The 2 wt.% Ni/$Ce_{0.7}Ti_{0.3}O_{2-\delta}$ catalyst was identified to deliver the best catalytic activity and stability among all the ceria supports and Ni loadings. At 650°C, the 2 wt.% Ni over $Ce_{0.7}Ti_{0.3}O_{2-\delta}$ catalyst shows good conversions of 73% and 79% for CH_4 and CO_2, respectively. The product yields were 42% and 52% for H_2 and CO. Additionally, compared to other metals including Pd and Co, the Ni catalyst delivers a higher reactivity and a long-term stability (up to 50 h) during the DRM reaction on stream. The enhanced reactivity and stability of this catalyst can be attributed to the unique interaction between the Ni metal and $Ce_{0.7}Ti_{0.3}O_{2-\delta}$ support, the high BET surface area (26 $m^2\ g^{-1}$) and metal active sites.

This research work is financially supported by Wyoming Carbon Engineering Initiative from School of Energy Resources at University of Wyoming and NSF (grant number CHE1151846).

536639 Controlled Metal@Metal Oxide Core-Shell Structures for Selective Heterogeneous Catalysis

Abstract:

Metal@metal oxide core-shell catalysts have attracted significant attention as compared to conventional supported catalysts. These structures facilitate construction of promising heterogeneous catalysts due to: (i) the high concentration of catalytically unique metal/metal oxide interfacial sites; (ii) improved chemical and thermal stability of small metal nanoparticles (NPs) by encapsulation with the metal oxides[1]; and (iii) spatially regulated delivery of reactants to the interfacial sites[2]. However, the controlled synthesis of these materials is challenging due to the large lattice mismatch between metal and metal oxide crystal structures, and the poor control over the synthesis of reducible metal oxide shells (such as TiO_2 and CeO_2) at room temperature induced by the highly reactive nature of the precursors, which leads to excessively fast, uncontrolled hydrolysis. We discuss a two-step synthesis method, known as â€œseeded growthâ€[3], as a feasible approach to overcome these challenges. A layer of ligand or surfactant molecules is introduced on the surface of the metal cores prior to encapsulation with the metal oxide shell. The molecules are able to reduce the large interfacial energy, as well as, minimize the aggregation of the NPs in solution. Pd@TiO_2 is used as a probe system to develop a controlled methodology for synthesizing metal@metal-oxide catalytic structures. The parameters that affect Pd NPs synthesis (such as reaction temperature, the amount of reducing agent and surface ligand) are investigated to achieve controlled Pd NP size and distribution. We find that the ligand coverage on the surface of Pd NPs can be affected by different reaction conditions leading to changes in the core-shell structures. Controlled synthesis of TiO_2 porous shell structures is achieved by varying parameters that affect the sol-gel reaction of titanium alkoxide precursor molecules (TAPs). We show that catalytic structures containing Pd NPs encapsulated within a porous TiO_2 film with small pores lead to high hydrodeoxygenation (HDO) reaction selectivity, while maintaining high catalytic activity. Catalyst selectivity is found to depend on both the presence of Pd-TiO_2 interfacial sites and the pore size of the TiO_2 shell[2]. The insights obtained from this systematic study provide a platform for controlled synthesis of metal@metal oxide core-shell structures with varying metal and metal oxide compositions.

REFERENCES

[1] Gould, T. D., Izar, A., Weimer, A. W., Falconer, J. L. and Medlin, J. W. ACS Catal. 4 (2014), 2714.

[2] J. Zhang, B. Wang, E. Nikolla, and J. W. Medlin, Angew. Chem. Int. Ed. 56 (2017), 6594.

[3] Li, G. & Tang, Z. Nanoscale 6 (2014), 3995.

531959 Protecting the Fe Active Phase from Oxidation Under Hydrodeoxygenation Conditions: Evaluating the Influence of Promoters and External Electric Fields

Abstract:

Bimetallic, promoted iron catalysts have demonstrated synergistic behavior that contributes to more cost-effective and longer-lasting hydrodeoxygenation (HDO) catalysts.[1-4] While precious metal promoters exhibit an overall synergistic interaction with Fe catalysts, further investigation into the behavior of oxygen on a promoted Fe-based catalytic grain is required to minimize the use of costly precious metals while maximizing the synergistic properties.[1, 2, 5] Furthermore, the application of externally applied electric fields has demonstrated favorable effects on unpromoted iron catalytic grains[6, 7] along with other catalytic processes, such as methane steam reforming.[8-10] Here, we use density functional theory and multi-scale modeling techniques to elucidate the effect of precious metal promoters and external electric fields on the kinetic behavior of oxygen over a full, multi-faceted catalytic Fe grain upon exposure to O_2 (Figure 1). Such a realistic, multi-faceted model of a promoted Fe grain is an effective tool for optimizing the use of precious metal promoters like Pd and Rh, taking advantage of an electric field as a means to fine-tune catalytic performance. These realistic, predictive models help to bridge the gap between experiment and theory, streamlining research efforts and accelerating progress.

In order to build the multi-faceted, promoted catalytic Fe grain model, the interaction of oxygen is calculated on both Pd and Rh promoted Fe(100), Fe(110), and Fe(111) over a range of oxygen surface coverages and electric field strengths (Figure 1). The data from the three Fe-promoted facets is then used to construct a model of a multi-faceted grain surface, incorporating the anisotropic effects of the surface orientation on its kinetic behavior.[11] Such a model accounts for the effect of pressure, temperature, electric field, promoters, and surface orientation on the adsorption/desorption and surface coverage behavior of oxygen. Preliminary results indicate that an Fe grain promoted with ¼ monolayer (ML) of either Pd or Rh experiences a massive reduction in oxygen surface coverage compared to the unpromoted system (Figure 1c). Moreover, the application of an electric field further reduces oxygen coverage but to a lesser degree than the promoter, consistent with adsorption energy trends in Figure 1b. The promoted Fe catalytic grain model can be used as an experimentally predictive, useful tool for optimizing methods in which HDO catalysts are designed to reduce oxidative deactivation while minimizing the use of costly precious metals.

Figure 1: The effect on ¼ ML Pd, ¼ ML Rh, and an applied electric field on oxygen's coverage distribution over a multi-faceted Fe grain, where (a) is a side view of the multi-faceted grain, (b) details the electric field effect of Pd and Rh on the adsorption energy of oxygen on Fe(110), and (c) is a top-down view of the oxygen coverage distribution over the multi-faceted grain surface, modeled at 650 K and 1×10^{-21} Pa.

References

1. Hong, Y., H. Zhang, J. Sun, K.M. Ayman, A.J.R. Hensley, M. Gu, M.H. Engelhard, J.-S. McEwen, and Y. Wang, *Synergistic Catalysis Between Pd and Fe in Gas Phase Hydrodeoxygenation of m-Cresol.* ACS Catalysis, 2014. **4**(10): p. 3335-3345.

2. Sun, J., A.M. Karim, H. Zhang, L. Kovarik, X.S. Li, A.J.R. Hensley, J.-S. McEwen, and Y. Wang, *Carbon-supported Bimetallic Pd–Fe Catalysts for Vapor-phase Hydrodeoxygenation of Guaiacol.* Journal of Catalysis, 2013. **306**: p. 47-57.

3. Sitthisa, S., W. An, and D.E. Resasco, *Selective Conversion of Furfural to Methylfuran Over Silica-Supported NiFe Bimetallic Catalysts.* Journal of Catalysis, 2011. **284**(1): p. 90-101.

4. Nie, L., P.M. de Souza, F.B. Noronha, W. An, T. Sooknoi, and D.E. Resasco, *Selective Conversion of m-Cresol to Toluene Over Bmetallic Ni–Fe Catalysts.* Journal of Molecular Catalysis A: Chemical, 2014. **388-389**: p. 47-55.

5. Hensley, A.J.R., Y. Hong, R. Zhang, H. Zhang, J. Sun, Y. Wang, and J.-S. McEwen, *Enhanced Fe2O3 Reducibility via Surface Modification with Pd: Characterizing the Synergy within Pd/Fe Catalysts for Hydrodeoxygenation Reactions.* ACS Catalysis, 2014. **4**(10): p. 3381-3392.

6. Bray, J., A.J.R. Hensley, G. Collinge, F. Che, Y. Wang, and J.-S. McEwen, *Modeling the Adsorbate Coverage Distribution Over a Multi-Faceted Catalytic Grain in the Presence of an Electric Field: O/Fe from First Principles.* Catalysis Today, 2018. http://dx.doi.org/10.1016/j.cattod.2018.04.016

7. Bray, J., G. Collinge, C. Stampfl, Y. Wang, and J.-S. McEwen, *Predicting the Electric Field Effect on the Lateral Interactions Between Adsorbates: O/Fe(100) from Frist Principles.* Topics in Catalysis, 2018. http://dx.doi.org/10.1007/s11244-018-0944-z

8. Che, F., J.T. Gray, S. Ha, and J.-S. McEwen, *Catalytic Water Dehydrogenation and Formation on Nickel: Dual Path Mechanism in High Electric Fields.* Journal of Catalysis, 2015. **332**: p. 187-200.

9. Che, F., J.T. Gray, S. Ha, and J.-S. McEwen, *Improving Ni Catalysts Using Electric Fields: A DFT and Experimental Study of the Methane Steam Reforming Reaction.* ACS Catalysis, 2016. **7**(1): p. 551-562.

10. Che, F., S. Ha, and J.-S. McEwen, *Hydrogen Oxidation and Water Dissociation over an Oxygen-Enriched Ni/YSZ Electrode in the Presence of an Electric Field: A First-Principles-Based Microkinetic Model.* Industrial & Engineering Chemistry Research, 2017. **56**(5): p. 1201-1213.

11. McEwen, J.-S., P. Gaspard, T.V. de Bocarmé, and N. Kruse, *Oscillations and Bistability in the Catalytic Formation of Water on Rhodium in High Electric Fields.* Journal of Physical Chemistry C, 2009. **113**(39): p. 17045-17058.

534412 Synthesis and Catalytic Testing of Lewis Acidic Nano-MFI Zeolites for Epoxide Ring Opening Reaction with Alcohol

Abstract:

The well-defined pores in zeolites impart shape selectivity for a variety of reactions. However, these pores can also cause diffusion limitations reducing the catalytic performance. Several strategies have been investigated for overcoming diffusion limitation including development of hierarchical structures and reduction in particle size. Nonetheless, synthesis of Lewis acidic nano-zeolites still remains a challenge. In this work, five-fold reduction in particle size is achieved for tin-substituted MFI (nSnMFI) zeolite by modifying the synthesis conditions, as compared to convention (cSnMFI). The growth of the nSnMFI is monitored using dynamic light scattering, showing that inclusion of Sn at 1:200 (Sn:Si) in the zeolite synthesis gel does not influence the zeolite growth rate. The increased catalytic performance because of enhanced diffusion in nSnMFI is demonstrated for epoxide ring opening reaction. Higher conversion is achieved for the ring opening of bulky epoxide substrate such as epoxyhexane with methanol with nSnMFI as compared to cSnMFI. The precise engineering of zeolites with reduced crystal size provides an efficient route to overcome diffusion limitations of bulky reactants. To further expand the scope, nano-zeolites with other frameworks can be achieved by modification of synthesis strategies.

529884 Understanding Intramolecular Cooperativity in Acid-Base Silica-Supported Organocatalysts

Abstract:

Carbon-carbon coupling reactions are essential steps in many chemical syntheses including pharmaceutical production and biomass valorization. Highly selective carbon-carbon bond formation steps are commonly performed with expensive homogeneous transition metal-complex catalysts, which suffer from process and scaling limitations. Heterogeneous organocatalysts, a class of enzyme-inspired catalysts which may be synthetically tuned and can incorporate multiple functionalities, are being investigated as possible immobilized and robust alternatives for various transformations. Weakly acidic mesoporous silica may be readily grafted with functional groups such as amines to form acid-base catalysts. Motivated by evidence that intramolecular cooperativity in bifunctional chemical moieties could provide additional benefit over catalysts with solely intermolecular cooperative interactions between grafted and surface species, this project studies how the alkyl linker length between the amine and alcohol moieties of the grafted alkanolamine silanes affect their intramolecular cooperativity. In order to isolate this activity, a parallel series of catalysts were synthesized with HMDS-capped surface silanols, removing the potential for intermolecular cooperativity between grafted amines and surface silanols. Kinetic studies of the acid-base catalyzed aldol condensation of 4-nitrobenzaldehyde and acetone were performed to determine relative catalyst activities and site-specific rates. Alkanolamine catalyst activity showed minimal dependence on alkyl linker length beyond 2 carbons, with comparable rates across 3-5 carbon spacings. Capped catalysts with solely intramolecular cooperative capabilities, however, were found to have activities strongly dependent on linker length. The capped 5 carbon linker catalyst was at least as active as the best uncapped catalyst, indicating effective catalytic cooperativity with a single bifunctional moiety.

Study of ethanol decomposition mechanism over combustion synthesized bimetallic Cu-Co nanoparticles

Anchu Ashok[1], Anand Kumar[2], Faris Tarlochan[1]

1. Department of Mechanical and Industrial Engineering, Qatar University, Doha, Qatar
2. Department of Chemical Engineering, Qatar University, Doha, Qatar

Email address of presenting author: akumar@qu.edu.qa

Introduction

It is well known that bimetallic catalysts show distinct physical and chemical properties differing from their constituent monometallic catalyst [1-9]. In this study, we investigate CuCo bimetallic catalyst for low temperature ethanol dehydrogenation reaction. Transition metals have traditionally been used for C-C bond cleavage, making them efficient catalysts for hydrocarbon and oxygenated hydrocarbon reforming reactions [10-17]. Steam reforming, partial oxidation, dehydrogenation and decomposition are the main routes for the hydrogen generation from ethanol. We follow decomposition route for H_2 production along with other necessary byproducts. Our findings indicate that bimetallic CuCo catalyst show good performance during ethanol decomposition for hydrogen production in the temperature range of 50°C - 400°C.

Nanomaterials are characterized with dimensions between 1-100 nm and are known to exhibit exceptional catalytic, optical and electronic properties making them suitable for a wide range of application [8,9]. Well controlled synthesis of nanoparticles requires time consuming and complicated steps involving bulky ligands as shape guiding agents and further post synthesis separation and purification. A simpler approach that has emerged over the past few years, known as combustion synthesis (CS), has gained considerable research attention on account of being economic, straight forward and simple synthesis procedure requiring considerably less synthesis time and exhibiting potential for synthesizing complex nanostructures [18]. A general form of redox reaction during combustion synthesis is expected to be as represented by the following equation [19]:

$$M^{v}(NO_3)_v + \left(\frac{5}{9}v\varphi\right) NH_2CH_2COOH + v\frac{5}{4}(\varphi - 1)O_2$$

$$\downarrow$$

$$M^{v}O_{\frac{v}{2}}(s) + \left(\frac{10}{9}v\varphi\right)CO_2(g) + \frac{25}{18}\varphi H_2O(g) + v\left(\frac{5\varphi + 9}{18}\right)N_2(g) \quad \text{--- (1)}$$

Where M^{v} is a v-valent metal. The parameter φ, known as fuel to oxidizer ratio, is defined such that $\varphi = 1$ corresponds to a stoichiometric oxygen concentration, meaning that the initial mixture does not require atmospheric oxygen for complete oxidation of the fuel, while $\varphi > 1$ (<1) implies fuel-rich (or lean) conditions.

In this paper we report the synthesis of Cu-Co catalyst nanoparticles using "Solution Combustion Synthesis (SCS)" technique. In this technique, an aqueous solution of metal nitrate and a reducing agent is homogeneously mixed and heated over a hot plate heater. After sometime self ignition leads to the start of combustion reaction that is self-sustained and produces nanoparticles with high porous structure. We have synthesized various transition metals and metal oxides, including bimetals and mixed metal-oxides using this technique [10-14,16,17,20-29]. This paper contains the description of synthesis of CuCo bimetal and oxides that are used for hydrogen production from ethanol dehydrogenation reaction.

Synthesis Procedure

Solution combustion synthesis (SCS) method is used to prepare bimetallic CuCo from the aqueous solution of copper nitrate ($Cu(NO_3)_2{\cdot}6H_2O$), cobalt nitrate ($Co(NO_3)_2{\cdot}6H_2O$), and glycine ($C_2H_5NO_2$) as fuel with a glycine to metal nitrate ratio of φ=0.5. An aqueous homogenous solution containing synthesis precursors in appropriate amount is heated over a hot plate heater to initiate the combustion reaction generating CuCo nanopowders. The resulted powder is ground using a mortar and pestle to get a uniform powder that was used to make catalyst pellets of size 0.6-1mm. Ethanol decomposition over bimetallic CuCo were conducted using in situ diffuse reflectance infrared fourier transform spectroscopy (DRIFT) study under N_2 flow at different temperatures (50, 100, 200, 300, 400°C). The synthesis conditions are well known to affect the particle morphology in SCS techniques. Amount of fuel and type of functional groups in fuels are reported to have significant impact on the metallic nature of the final product (metal vs. metal oxide) in addition to crystallite size, porosity, surface structure and the composition of carbon in the final product [23,24,29]. The physio-chemical characteristics of nanoparticles and their catalytic efficiency are interconnected, and controlling physio-chemical properties can greatly enhance catalytic performance [7,9,23,24,29]. In this section we present the details of synthesis of Cu and Co based catalysts using SCS technique for catalytic hydrogen generation from ethanol.

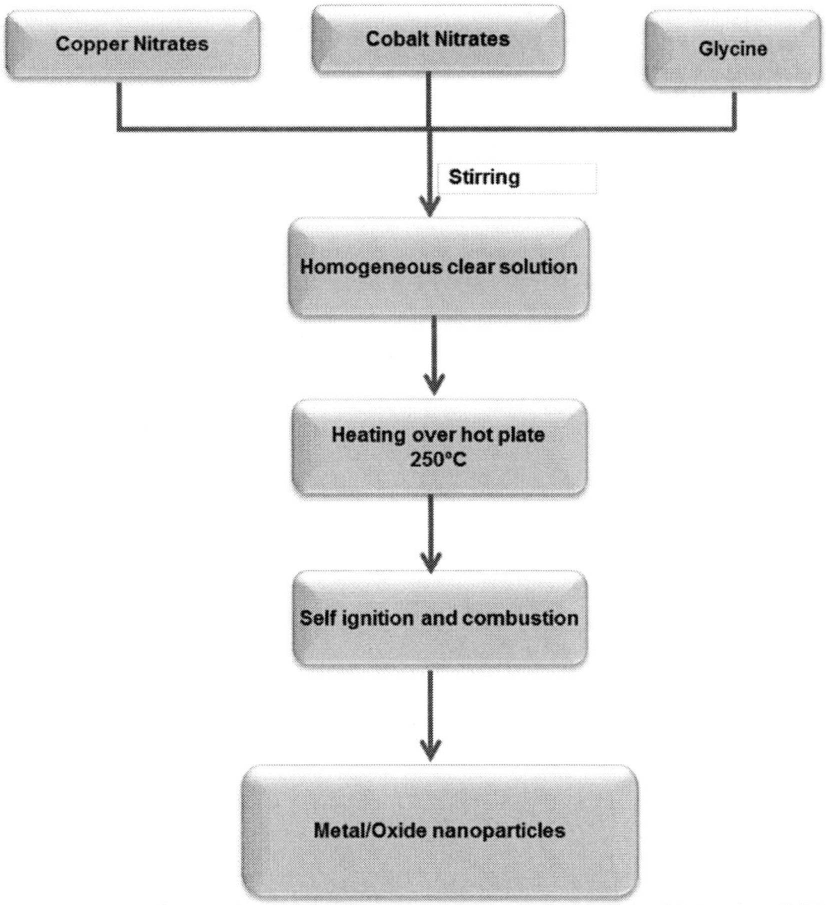

Figure 1: Flow chart shows the method of synthesis of bimetal (oxide) using SCS method

The flow chart (Fig. 1) shows the steps involved in the method of synthesis of metals/metal-oxides using solution combustion synthesis. In this case the homogeneous solution consisting of copper nitrate and cobalt nitrate (1:1 molar ratio) is prepared and a predefined amount of glycine as a fuel (phi = 0.5) is added. This solution is stirred until a clear homogeneous and diluted solution is obtained, which is placed on a hotplate heater maintained at 250°C temperature. This solution gradually loses water and becomes thick and soapy before suddenly combusting in air. The remains after the combustion contains particle with size in the range of nanometers [25,26]. The temperature-time profile for the metal synthesis are recorded (as deemed necessary) using a Daq board (and data logger) to collect information about the ignition temperature along with the reduction reaction temperature of the solution. It also provides a reasonable indication whether the synthesized product will be in metallic or oxide form. The as-synthesized particles were characterized to study its morphology, particle size. Surface area of the particles were carried out on BET Micromertics ASAP 2420 in the absorbent median as nitrogen gas The sample were exposed to N_2, after it attain a differential pressure below 20 μm by outgassed it through a temperature of 473K followed by immersing in liquid nitrogen. XRD characterization were done using Rigaku MiniFlexII Desktop-ray powder diffractometer with a wavelength of 1.54056 A^0 by Cu-Kα radiation with a scan range of 10-90 deg. Crystalline size can be determined with the identified peaks using Scherrer equation [30] as:

$$\tau = \frac{k\lambda}{\beta cos\theta} \qquad \text{---------------------- (2)}$$

Where k is the shape factor, with value close to unity, λ be the x-ray wavelength β is the line broadening at half maximum intensity and θ is the angle of diffraction.

Morphology analysis were carried out with Scanning Electron Microscope(SEM) Nova Nano 450 (FEI), featured with non-destructive, high resolution image reaches a magnification up to 2500 kx with the highly focused scanning electron beam. An Energy-dispersive X-ray spectroscopy (EDX), Bruker 127eV coupled with the SEM is capable for the quantitative analysis, entails the measuring the elemental composition present in the sample. Such a quantitative study allows the volume analysis of the sample as small as 1 micron. Finally, Powdered nanocompounds were characterized using High Resolution Transmission electron microscopy (HR TEM), TECNAI G^2 F20 High Vacuum and Resolution Field Emission with flexible and continuous high tension power ranging between 20 to 200 KV and consisting of a specimen holder placed in the liquid nitrogen chamber, providing the instrument to measure the individual molecules and its size in very high resolution. To avoid the agglomeration in the sample, it has been preferred to do ultrasonic agitate after dispersing the sample with ethanol. Only selected characterization results are discussed in this paper, more details will be presented during the conference and future communications.

DRIFTS studies were conducted on Thermo Nicolet 6700 FTIR containing praying mantis mirror assembly and a DRIFTS reactor (both from Harrick Scientific). The synthesized nanopowders were hand ground using a mortar and pestle and screened to get particles size < 106μm. These powders were placed inside DRIFTS reactor sample holder and lightly pressed to get a flat surface. Before conducting ethanol dehydrogenation mechanist studies, the catalysts were activated using a hydrogen flow (100ccm) at 300°C for one hour, thereafter the flow was switched to nitrogen (100ccm) to purge the system. Background spectra were collected at different temperatures under nitrogen flow and the reactor was cooled down to 50°C to start the DRIFTS experiments. Ethanol was introduced to the reactor by passing nitrogen stream through a bubbler containing ethanol to sweep ethanol vapors to the reactor. Temperature was increased from 50°C to 400°C and sample spectra were collected at desired temperature values. The spent catalysts after reaction were collected and characterized for comparison purpose.

Results and Discussion

The copper-cobalt catalyst was synthesized using the technique described in the previous section. The presence of bimetallic copper-cobalt and their oxidized states were confirmed using XRD. SEM study indicates the agglomerated nanoparticles of high porosity that are randomly distributed. EDS and XRD results indicates the presence of higher amount of oxide in the synthesized sample indicating the form to be like $CuCoO_2$. SEM image in Fig. 2 also shows the non-uniform sized particles of bimetallic CuCo along with agglomeration that are common in solution combustion mode. Particle size is in the range of 20-30 nm. Copper-cobalt oxide prepared using combustion technique was reduced prior to the decomposition reaction by passing H_2 in the reaction chamber at 300°C. Individual metals are also used for DRIFTS studies to understand the role of each element on ethanol decomposition. Only results on copper surface is described here, other results including cobalt and copper-cobalt bimetallic-oxide will be communicated in future. Fig. 3 summarizes the copper results that is based on the interpretation of the FTIR data to provide reaction pathway of ethanol decomposition on copper surface. The presence of adsorbed ethanol and ethoxy species over the reduced catalyst was clear from the FTIR spectrum between 50-100°C. Changes in IR peaks were followed with temperature to decipher the reaction pathway. IR peak at 3669cm^{-1} shows the clear presence of adsorbed OH on the catalyst surface. These results indicate bonding to be from OH site that further is transforms ethanol into ethoxy intermediate. Above 150°C, adsorbed ethoxy species lose hydrogen from CH and indicate the formation of CO bond as seen in aldehyde compounds. Increasing the temperature futher leads to gaseous acetaldehyde. Copper-cobalt, on the other hand, displays slight change in reaction pathway after ethoxy species generation. At 150°C, ethoxy intermediates are converted to acetate species along with carbonate compound. Increasing the temperature above 200ºC intensifies the peak at 2335-2367 cm^{-1} showing formation CO_2 during the reaction. At that period, the acetate species are possibly dehydrogenated to acetaldehyde and other intermediate species. Some carbon formation was also observed at higher temperature.

Fig 2. SEM of bi-metallic (oxide) of Cu and Co prepared using SCS

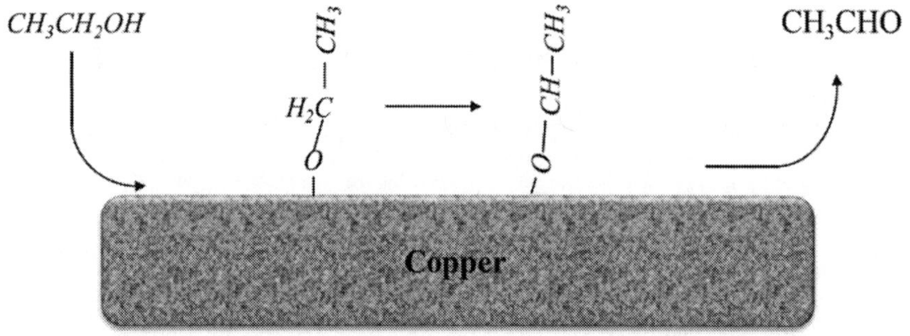

Fig 3. Reaction pathway on copper surface based on DRIFTS analysis and interpretations

Conclusion

Solution combustion synthesis was used to synthesize metals/oxide of copper-cobalt. Based on our experimental analysis, the SCS method is suitable for synthesizing pure metal oxides, with high porosity and surface area suitable for catalytic applications. DRIFTS studies on copper, cobalt and copper-cobalt bimetallic oxide was conducted to understand the ethanol decomposition reaction pathway on the three surfaces. Copper is found to be selective for acetaldehyde formation, whereas cobalt and copper-cobalt bimetallic oxides are found to generate intermediate carbonate species that decompose at elevated temperatures to produce carbon dioxide, acetaldehyde and possibly methane. Further work related to more detailed characterizations (e.g. TEM, elemental phase mapping and XPS) is underway and will be included in future correspondences.

Acknowledgement:

This publication was made possible by NPRP grant (NPRP8-509-2-209) from the Qatar National Research Fund (a member of Qatar Foundation). The statements made herein are solely the responsibility of the author(s).

References:

[1] B. Han, Y. Yang, Y. Xu, U. Etim, K. Qiao, B. Xu, Z. Yan, A review of the direct oxidation of methane to methanol, Chinese Journal of Catalysis. 37 (2016) 1206-1215.

[2] T.M. Gür, Comprehensive review of methane conversion in solid oxide fuel cells: prospects for efficient electricity generation from natural gas, Progress in Energy and Combustion Science. 54 (2016) 1-64.

[3] D. Pakhare, J. Spivey, A review of dry (CO 2) reforming of methane over noble metal catalysts, Chem. Soc. Rev. 43 (2014) 7813-7837.

[4] S. Dutta, A review on production, storage of hydrogen and its utilization as an energy resource, Journal of Industrial and Engineering Chemistry. 20 (2014) 1148-1156.

[5] Y.T. Shah, T.H. Gardner, Dry Reforming of Hydrocarbon Feedstocks, Catalysis Reviews. 56 (2014) 476-536.

[6] Y. Izumi, Recent advances in the photocatalytic conversion of carbon dioxide to fuels with water and/or hydrogen using solar energy and beyond, Coord. Chem. Rev. 257 (2013) 171-186.

[7] Z. Wei, J. Sun, Y. Li, A.K. Datye, Y. Wang, Bimetallic catalysts for hydrogen generation, Chem. Soc. Rev. (2012).

[8] F. Tao, Synthesis, catalysis, surface chemistry and structure of bimetallic nanocatalysts, Chem. Soc. Rev. 41 (2012) 7977-7979.

[9] W. Yu, M.D. Porosoff, J.G. Chen, Review of Pt-Based Bimetallic Catalysis: From Model Surfaces to Supported Catalysts, Chem. Rev. 112 (2012) 5780-5817.

[10] A. Ashok, A. Kumar, R. Bhosale, F. Almomani, M.A.H.S. Saad, S. Suslov, F. Tarlochan, Influence of fuel ratio on the performance of combustion synthesized bifunctional cobalt oxide catalysts for fuel cell application, in press. (2018).

[11] A. Ashok, A. Kumar, R. Bhosale, M.A. Saleh Saad, F. AlMomani, F. Tarlochan, Study of ethanol dehydrogenation reaction mechanism for hydrogen production on combustion synthesized cobalt catalyst, International Journal of Hydrogen Energy. 42 (2017) 23464-23473.

[12] A. Kumar, A. Ashok, R.R. Bhosale, M.A.H. Saleh, F.A. Almomani, M. Al-Marri, M.M. Khader, F. Tarlochan, In situ DRIFTS Studies on Cu, Ni and CuNi catalysts for Ethanol Decomposition Reaction, Catal. Lett. 146 (2016) 778-787.

[13] A. Kumar, A. Cross, K. Manukyan, R.R. Bhosale, L.J.P. van den Broeke, J.T. Miller, A.S. Mukasyan, E.E. Wolf, Combustion synthesis of copper–nickel catalysts for hydrogen production from ethanol, Chem. Eng. J. 278 (2015) 46-54.

[14] A. Kumar, J.T. Miller, A.S. Mukasyan, E.E. Wolf, In situ XAS and FTIR studies of a multi-component Ni/Fe/Cu catalyst for hydrogen production from ethanol, Applied Catalysis A: General. 467 (2013) 593-603.

[15] G. Carotenuto, A. Kumar, J. Miller, A. Mukasyan, E. Santacesaria, E. Wolf, Hydrogen production by ethanol decomposition and partial oxidation over copper/copper-chromite based catalysts prepared by combustion synthesis, Catalysis Today. 203 (2013) 163-173.

[16] A. Cross, A. Kumar, E.E. Wolf, A.S. Mukasyan, Combustion Synthesis of a Nickel Supported Catalyst: Effect of Metal Distribution on the Activity during Ethanol Decomposition **2012**, *51* (37), pp 12004–12008, Ind Eng Chem Res. 51 (2012) 12004-12008.

[17] A. Kumar, A. Mukasyan, E. Wolf, Combustion synthesis of Ni, Fe and Cu multi-component catalysts for hydrogen production from ethanol reforming, Applied Catalysis A: General. 401 (2011) 20-28.

[18] A. S. Mukasyan, P. Dinka, Novel Method for Synthesis of Nano-Materials: Combustion of Active Impregnated Layers, Adv. Eng. Mater. 9 (2007) 653-657.

[19] P. Dinka, A.S. Mukasyan, In Situ Preparation of Oxide Based Supported Catalysts by Solution Combustion Synthesis, J. Phys. Chem. 109 (2005) 21627-21633.

[20] M.A. Matin, A. Kumar, Saad, Mohammed Ali H Saleh, M.J. Al-Marri, S. Suslov, Zn-enriched PtZn nanoparticle electrocatalysts synthesized by solution combustion for ethanol oxidation reaction in an alkaline medium, MRS Communications. 8 (2018) 411-419.

[21] A. Ashok, A. Kumar, R.R. Bhosale, F. Almomani, S.S. Malik, S. Suslov, F. Tarlochan, Combustion synthesis of bifunctional $LaMO_3$ (M=Cr, Mn, Fe, Co, Ni) perovskites for oxygen reduction and oxygen evolution reaction in alkaline media, J. Electroanal. Chem. 809 (2018) 22-30.

[22] M. Matin, A. Kumar, R. Bhosale, M.S. Saad, F. Almomani, M. Al-Marri, PdZn nanoparticle electrocatalysts synthesized by solution combustion for methanol oxidation reaction in an alkaline medium, RSC Advances. 7 (2017) 42709-42717.

[23] A. Ashok, A. Kumar, R.R. Bhosale, M.A.H. Saleh, U.K. Ghosh, M. Al-Marri, F.A. Almomani, M.M. Khader, F. Tarlochan, Cobalt oxide nanopowder synthesis using cellulose assisted combustion technique, Ceramics International. 42 (2016) 12771-12777.

[24] A. Ashok, A. Kumar, R.R. Bhosale, M.A.H. Saleh, van den Broeke, Leo JP, Cellulose assisted combustion synthesis of porous Cu–Ni nanopowders, RSC Advances. 5 (2015) 28703-28712.

[25] A. Kumar, E.E. Wolf, A.S. Mukasyan, Solution Combustion Synthesis of Metal Nanopowders: Nickel-Reaction Pathways, AIChE Journal. 57 (2011) 2207-2214.

[26] A. Kumar, Combustion Synthesis of Novel Catalysts for Hydrogen Production from Light Alcohols, University of Notre Dame, 2011.

[27] A. Kumar, E.E. Wolf, A.S. Mukasyan, Solution combustion synthesis of metal nanopowders: Copper and copper/nickel alloys, AIChE Journal. 57 (2011) 3473-3479.

[28] A. Kumar, A.S. Mukasyan, E.E. Wolf, Impregnated layer combustion synthesis method for preparation of multicomponent catalysts for the production of hydrogen from oxidative reforming of methanol, Applied catalysis.A, General. 372 (2010) 175-183.

[29] A. Kumar A.S. Mukasyan, E.E. Wolf, Modeling Impregnated Layer Combustion Synthesis of Catalyts for Hydrogen Generation from Oxidative Reforming of Methanol, Ind. Eng. Chem. Res. 49 (2010) 11001-11008.

[30] A.F. Ghenciu, Review of fuel processing catalsyts for hydrogen production in PEM fuel cell systems, Curr.Opin.Solid State Mater.Sci. 6 (2002) 389-399.

532797 Synergetic Effect of Ultrafine Nico Bimetallic Alloy Nanoparticles Derived from Bimetal-Organic Frameworks

Abstract:

Reducing the size of metal particles to the subnanometer-scale in catalysis is attractive, because metal clusters composed of several atoms exhibit novel properties that differ greatly from those predicted by simple scaling laws. In this study, highly dispersed NiCo bimetallic alloy nanoparticles have been successfully immobilized on the SiO_2 frameworks by using hetero-nuclear metal-organic frameworks (MOFs) as metal alloy precursors. Catalyst characterizations revealed that the average size of NiCo alloy particles was less than 1 nm, with a total metal loading of about 20 wt%. Compared with individual Ni or Co MOF-derived catalysts and the catalysts prepared by the conventional impregnation method, the ultrafine NiCo/SiO_2-MOF catalyst showed a much better catalytic performance in the catalytic hydrogenation of furfuryl alcohol (FA) to tetrahydrofurfuryl alcohol (THFA) under mild conditions, giving 99.8% conversion of FA and 99.1% selectivity to THFA. It was found that a significant synergistic effect existed between Co and Ni within the subnanometer NiCo/SiO_2-MOF catalyst, which was twice and 20 times more active than Ni/SiO_2-MOF and Co/SiO_2-MOF, respectively. In contrast, such a synergistic effect is not observed within the NiCo/SiO_2-IMP catalyst prepared by conventional impregnation. This interesting finding of size-dependent bimetallic synergetic effect provides guidance for enhancing the activity of bimetallic catalysts.

Figure 1. Reaction constants of different catalysts for FA hydrogenation to THFA.

536668 Synthesis of Carbon Quantum Dots from Ohio-Derived Coal

Abstract:

Abstract

While coal has been an energy source for many decades, its value has been reduced due to environmental concerns related to greenhouse gas emissions. Coal could represent a carbon source for many industrial chemicals. We focus on Ohio-derived coal as a raw material to generate carbon quantum dots (CQDs). Carbon quantum dots (CQDs) are biocompatible, possess unique optical properties, and possess application in a variety of emerging processes. [1]. The high surface area inherent to these materials makes them applicable to electrochemical capacitors as energy storage materials.

This presentation focuses on our efforts to synthesize carbon quantum dots from Ohio coal. We will explore safe, environmentally friendly routes to synthesize CQDs from coal, and will investigate their applicability as energy storage materials, especially as electrochemical capacitors.

References:

[1] S. Ying Lim, W. Shen, and Z. Gao, â€œCarbon quantum dots and their applications,â€ Chemical Society Reviews, vol. 44, no. 1, pp. 362â€“381, 2015.

538893 Template-Free Self-Assembly of 3D Graphene/Noble Metal Nanotube Composite Electrocatalysts for Oxygen Reduction Reaction in Fuel Cells

Abstract:

Template-Free Self-Assembly of 3D Graphene/Noble Metal Nanotube Electrocatalysts for Oxygen Reduction Reaction in PEM Fuel Cells

Enoch A. Nagelli*, Gabrielle M. Milanesa, Kamil Woronowicz, F. John Burpo, Alexander N. Mitropoulos

Department of Chemistry & Life Science

United States Military Academy, West Point, New York 10996

*Corresponding PI: Dr. Enoch Nagelli, Email: enoch.nagelli@usma.edu

Graphene is a promising nanoscale platform to construct next generation three-dimensional electrodes for high performance, light-weight, and configurable energy storage devices. However, the design, synthesis, and assembly of nanomaterials with well-defined structure and the connection from individual nanoscale entities to these real world applications continues to present big challenges. This is critical to build nanostructured electrodes to develop high performance batteries and fuel cells. In order to achieve this, we demonstrate a scalable method using a solution based electrostatic self-assembly technique for the formation of novel 3D noble metal micro- or nanotube composites with graphene. For example, we electrostatically bind graphene oxide with platinum salt coordination complexes and chemically reduce the resultant 3D structure using sodium borohydride. Our design process enables the controlled formation and assembly of micro- or nanotubes onto 2D graphene sheets in order to develop new high surface area composite electrodes with any pure noble metal or alloy noble metals. This novel method of aqueous assembly is a platform which can lead to research and development of nanocomposite materials and structures that will enable development of high performance electrocatalysts for oxygen reduction reaction.

References:

1. J. Burpo, E. A. Nagelli, S. J. Winter, J. P. McClure, S. F. Bartolucci, A. R. Burns, S. F. O'Brien, D. Chu, "Salt-Templated Hierarchically Porous Platinum Macrotube Synthesis" *ChemistrySelect* Accepted March **2018.**
2. Burpo, E. A. Nagelli, L. Morris, J. McClure, M. Ryu, J. Palmer, "Direct solution-based reduction synthesis of Au, Pd, and Pt aerogels" *J. Mater. Res.* **2017**, 22, 4153.
3. Nagelli, L. Huang, F. Du, and L. Dai, "3D Vertically-Aligned CNT/Graphene Hybrids from Layer-by-Layer Transfer for Supercapacitors" *Part. Part. Syst. Charct.* **2017**, 34, 1700131. Invited Article in Special Issue of Graphene Oxide Liquid Crystals.
4. Georgakilas, M. Otyepka, A. B. Bourlinos,V. Chandra, N. Kim, K. C. Kemp, P. Hobza, R. Zboril, K. S. Kim, "Functionalization of Graphene: Covalent and Non-Covalent Approaches, Derivatives and Applications" *Chem. Rev.* **2012**, *112*, 61564.

525964 Hydrogen Storage in Small PtPd Alloy Nanoparticles: A DFT Study

Abstract:

Metal hydrides are promising hydrogen storage materials as they can store large amounts of hydrogen in a small space (volumetric densities >100 kg H_2 m^{-3}). Transition metal hydrides are one of the few systems that exhibit the ability to reversibly absorb hydrogen near ambient conditions.[1,2] However, they are heavy and expensive. One possible way to reduce the amount of metal needed and/or increase the amount of hydrogen stored is to use smaller nanoparticles. Surprisingly however, Pd exhibits very different behavior compared to Pt when nanosized. While the hydrogen storage capacity of Pd decreases with decreasing size,[3] that of Pt in fact increases.[4] Insights on the mechanisms by which hydrogen is stored in metals are needed to help us understand the effects of nanosizing on their capacity.

To gain insights into the mechanisms by which hydrogen is stored in small nanoparticles, we performed density functional theory (DFT) calculations to study: *i)* the effect of the metal, *ii)* the effect of core-shell morphology, and *iii)* the effects of alloyed surfaces. At realistic hydrogen pressures, high coverages of hydrogen are expected on the nanoparticles. We thus studied the adsorption of multiple hydrogens on surface as well as subsurface sites. The most stable configurations of the different hydrogen coverages on the nanoparticle were obtained by an iterative process involving fitting of a Cluster Expansion (CE) Hamiltonian,[5,6] which parameterizes the DFT energies we calculated in terms of point and multibody interactions, and employing the CE Hamiltonian in a simulated annealing procedure to locate stable minima.

Our results show that surface sites can contribute significantly to the hydrogen storage capacity in small nanoparticles. This is as opposed to the commonly-held belief that hydrogen is only stored in the bulk.[3,4] The insights we have gained will serve as a guide for designing higher capacity, less costly, hydrogen storage materials in the nanosized regime.

References

1 L. Schlapbach and A. Züttel, *Nature*, 2001, **414**, 353–358.

2 A. Züttel, *Mater. Today*, 2003, **6**, 24–33.

3 M. Yamauchi, R. Ikeda, H. Kitagawa and M. Takata, *J. Phys. Chem. C*, 2008, **112**, 3294–3299.

4 M. Yamauchi, H. Kobayashi and H. Kitagawa, *ChemPhysChem*, 2009, **10**, 2566–2576.

5 J. M. Sanchez and D. De Fontaine, in *Structure and Bonding in Crystals*, eds. A. Nsvrotsky and M. O'Keeffe, Academic Press, Inc, New York, 1981, vol. 2, p. 117.

6 D. J. Schmidt, W. Chen, C. Wolverton and W. F. Schneider, *J. Chem. Theory Comput.*, 2012, **8**, 264–273.

538067 Salt-Ceramic Composite Electrolytes for Lithium Metal Batteries

Abstract:

Lithium metal based all solid state batteries are promising candidates for next generation energy storage system due to their high energy densities. A key challenge for lithium metal batteries is the development of a solid electrolyte because they have good chemical and thermal stability, robust mechanical strength, and are able to stop dendrite growth of Li anodes. However, present processing method of solid electrolytes requires high temperature to sinter ceramic powder, which limits opportunities for integration soft materials and leads to high processing costs. Here, we developed salt-ceramic composite electrolytes through a cold sintering process that utilizes a small amount of solvent and uniaxial pressure to sinter ceramics at low temperatures (~130 °C). By synthesizing salt-LATP composite electrolytes with various lithium salts, conductivities at room temperature close to 8×10^{-4} S/cm were achieved; this is one of the highest value obtained from LATP electrolytes so far. In addition, we also found out that a lithium salt was able to improve the sinterability of LATP, resulting in increased relative density of composite electrolytes. We believe that our approach of salt-ceramic composites through cold sintering may make an impact in a variety of fields, including solid electrolytes, composite cathodes, and tough or responsive structural materials.

534955 Atomic Layer Deposition of Nanoscale Solid State Electrolyte for the Next-Generation Energy Storage

Abstract:

Atomic Layer Deposition of Nanoscale Solid State Electrolyte for the Next-Generation Energy Storage

Chuan-Fu Lin and Gary Rubloff

Department of Materials Science and Engineering, University of Maryland, College Park, MD 20742

Over the past few years, atomic layer deposition (ALD) has emerged as a promising technique to develop electrochemical active materials that can be effectively coated on random 3D nanostructures and to improve the performance of electrochemical energy storage devices. In conventional battery architectures and chemistries, ALD layers have been used to protect the surface of anodes and cathodes, resulting in improved electrode stability and enhanced cycling performance. Beyond conventional Li-ion batteries, ALD has allowed fabrication of ultra-high performance 3D nanoheterostructures that combine intimate contact between high electrical conductivity current collectors and thin layers of active material for incredible charge/discharge rate capabilities proportional to the active material surface area. Both anode and cathode nanoheterostructures exhibit significantly higher charge/discharge rate capabilities over their bulk electrode counterparts in ½ cell geometries. The culmination of this architecture uses the high degree of control inherent in the ALD process to fabricate arrays of opposing cylindrical heterostructures inside a nanoporous template, and can achieve nearly 50% theoretical capacity while charging/discharging at a rate of 150 C.

However, use of liquid electrolytes present a number of challenges with advance, next-generation high power electrodes. High charge/discharge rates, desirable for fast charging of plug-in hybrid vehicles and personal electronic devices exacerbate already present safety issues, particularly that of Li dendrite formation and subsequent cell failure. Increased electrode surface area may mitigate dendrite formation by lowering the local current density, but deleterious electrolyte degradation reactions can consume electrolyte and result in eventual cell failure. In this work, we demonstrate that ALDâ€™s capability for very thin solid electrolyte layers opens the door to advanced electrode protection, and recent results showcasing the utility of ALD solid electrolyte coatings for protection layers on 3D nanoheterostructured electrodes and next-generation Li metal anodes will be discussed. In the case of conversion electrode materials, ALD coatings can prevent electrolyte decomposition and solid electrolyte interphase formation, lowering overpotential required for charge and discharge. Additionally, ALD coatings mechanically constrain conversion electrode particles, preventing volume expansion and fracturing of the coatings upon cycling, and surprisingly suppress the irreversible phase transformation. ALD layers applied to Li metal anodes can prevent electrolyte decomposition and Li dendrite formation by chemical stabilization of the electrode surface, resulting in both higher capacities and longer cycling lifetimes in the case of the Li-S and Li-Air systems.

The vapor phase synthesis methods of ALD provide large opportunity to synthesize broad materials chemistry of our interests, and here in this work, we will introduce our efforts on synthesize Li-containing materials by adding different functionalities and reaction chemistry. We explored the development and the materials chemistry of ALD inorganic solid electrolyte, organic-inorganic (hybrid-flexible) solid electrolyte, and so on.

Use of solid state electrolytes, previously only suitable for niche applications due to their low inherent ionic conductivity and limitations in fabrication process, can now be expanded to new applications such as the previously described 3D batteries by utilizing ALD processes. A number of suitable ALD process for well-known solid electrolytes have been developed allowing deposition of thin, conformal solid electrolyte layers. The current status of 3D solid state batteries will be discussed, as will remaining challenges.

538295 Molecular Engineering of Hydroxide Ion Conducting Aromatic Polymers and Their Applications in Alkaline Membrane Fuel Cells

Abstract:

Molecular Engineering of Hydroxide Ion Conducting Aromatic Polymers and Their Applications in Alkaline Membrane Fuel Cells

Junyoung Han,[†] Jong Yeob Jeon,[†] Sangtaik Noh,[†] Chang Y. Ryu,[†] Chulsung Bae*,[†§]

[†]Department of Chemistry and Chemical Biology, [§] Department of Chemical and Biological Engineering, Rensselaer Polytechnic Institute, Troy, New York 12180, United States

Recently, anion exchange membrane (AEM) fuel cell has attracted attention as a clean energy technology device to convert chemical energy into electric energy. AEM fuel cells have several advantages, such as fast kinetics and non-precious metal catalysts.[1] However, AEMs that have desired properties (e.g. high hydroxide conductivity, thermochemical stability at high pH for durable operation, and convenient synthetic process) are not commercially available yet.[2,3] New material developments by designing of new polymer backbones, side chain functional groups and new anion conductive groups have been extensively investigated over the past decade.[4] Several research groups including us have reported that a long side chain between polymer backbone and quaternary ammonium (QA) end group is helpful for enhancing micro-phase separation based on thermochemical stable backbone.[5-7]

In this study, we will present the effects of polymer structures on the membrane morphology, properties and fuel cell performance. To better understand polymer structure–membrane property relationships, aromatic polymers with different QA head groups, tether chain lengths, polymer backbone were prepared and their properties were evaluated. In addition to the fundamental study, how high performance AEMs can be produced in a kilo gram scale by cost effective processes will be presented.

References

1. Y. J. Wang, J. Qiao, R. Baker, J. Zhang, J., *Chem. Soc. Rev. 42*, 5768, (2013).
2. O. M. M. Page, S. D. Poynton, S. Murphy, A. Lien Ong, D. M. Hillman, C. A. Hancock, M. G. Hale, D. C. Apperley, J. R. Varcoe, *RSC Adv. 3*, 579 (2013).
3. C. Fujimoto, D.-S. Kim, M. Hibbs, D. Wrobleski, Y. S. Kim, *J. Membr. Sci. 423–424*, 438 (2012).
4. A. D. Mohanty, C. Y. Ryu, Y. S. Kim, C. Bae, *Macromolecules 48*, 7085 (2015).
5. H.-S. Dang, P. Jannasch, *Macromolecules 48*, 5742 (2015).
6. W.-H. Lee, Y. S. Kim, C. Bae, *ACS Macro Lett.* **4**, 814 (2015).
7. W.-H. Lee, A.D. Mohanty, C. Bae, *ACS Macro Lett.* **4**, 453 (2015).

522128 Computational Design of Electrochemical CO_2 Reduction Catalysts

Abstract:

The importance of electrochemical catalysts rapidly growing due to their wide applications in renewable energy conversion and storage. Based on the fundamental mechanism of the electrochemical catalysts, to understand, predict, and thereby design the electrochemical catalyst materials, we develop first-principles based multi-scale computational methods. In this talk, I will discuss our recent works demonstrating how the computational materials simulations can aid understanding the electrochemical systems and designing efficient catalyst materials for the CO_2 conversion technology, which is considered as a complementary part of hydrogen technology. On the basis of our series works, we suggest design principles to realize efficient CO_2 conversion catalyst materials. We expect this work to provide useful insights to guide the development of a feasible strategy for CO_2 conversion, which will help developing a sustainable economy for the present and the future.

[1] Hyung-Kyu Lim, Hyeyoung Shin, *et al.*, *J. Am. Chem. Soc.*, **2014**, 136, 11355

[2] Cheonghee Kim, *et al.*, *J. Am. Chem. Soc.*, **2015**, 137, 13844

[3] Ki Dong Yang, Yoonhoo Ha, *et al.*, *Adv. Funct. Mater.*, **2016**, 26, 233

[4] Kang-Shan Kim, *et al.*, *ACS Catal.*, **2016**, 6, 4443

[5] Da Hye Won, Hyeyoung Shin, *et al.*, *Angew. Chem. Int. Ed.*, **2016**, 55, 9297

[6] Hyeyoung Shin, *et al.*, *J. Phys. Chem. Lett.*, **2016**, 7, 4124

[7] Cheonghee Kim, Taedaehyeoung Eom *et al.*, *ACS Catal.*, **2017**, 7, 779

[8] Jai Hyun Koh, Da Hye Won, Taedaehyeoung Eom *et al.*, *ACS Catal.*, **2017**, 7, 5071

[9] Hyung-Kyoo Lim *et al.*, *ACS Catal.*, **2018**, 8, 2420

[10] Taedaehyeong Eom *et al.*, *J. Phys. Chem. C*, **2018**, DOI: 10.1021/acs.jpcc.8b02886

522138 Ni-Fe Alloy Nanowire Arrays As Outstanding Bifunctional Electrocatalysts for Overall Water Splitting

Abstract:

Renewable energy driven electrocatalytic water splitting has been considered by many as the most promising green hydrogen production technology. Its high electricity consumption cost however prevents it from becoming the prevailing hydrogen production technology. Highly efficient and durable low cost electrocatalysts are the key to reduce its production cost to make it more competitive with other currently significantly cheaper hydrogen production technologies. Herein, we develop FeNi alloy nanowire arrays as outstanding bifunctional electrocatalysts for overall water splitting. The overpotentials of Ni nanowire arrays for the hydrogen evolution reaction at current densities of 20 and 400 mA/cm^2 are only 112 and 346 mV, respectively, and those of FeNi-nanowire arrays for the oxygen evolution reaction at current densities of 20 and 400 mA/cm^2 are only 249 and 326 mV, respectively. The overall water splitting is conducted by using Ni nanowire arrays as the cathode and NiFe nanowire arrays as the anode, achieving 10 mA/cm^2 at an ultralow potential of 1.61 V. The overall water splitting efficiency remains nearly unchanged after 24 hours operation at a high current density of 400 mA/cm^2. The outstanding stability and low overpotentials at high current densities demonstrates that the present electrocatalysts are promising for large scale practical applications.

527665 Rational Design of Single-Atom Electrocatalysts for Hydrogen Evolution Reaction

Abstract:

Hydrogen is regarded as a renewable and clean fuel in the future energy infrastructure. In order to produce hydrogen, water splitting through electrochemical hydrogen evolution reaction (HER) in acid media is one of the most effective technologies. However, since the most common electrode materials are the Pt-based catalysts, tremendous efforts have been devoted toward reducing their costs. Single-atom catalysts (SACs) have aroused increasing interest for practical applications due to the low cost and the improvement of their intrinsic performance. In this talk, computational design of SACs for HER will be discussed. The potential supports such as TiC and WO_3 for SACs were investigated by using density functional theory calculations. Specifically, we found a suitable reaction condition to stabilize SACs and explored the reaction mechanism of HER on the SACs. The strategy developed in this study may have wide applications to various heterogeneous catalysis systems, including SAC for other applications

529538 Nano-Structure Analysis of Catalyst Layer in Polymer Electrolyte Fuel Cell

Abstract:

Fuel cell electric vehicle (FCEV) is an effective solution to reduce the consumption of petroleum fuel and the emission of carbon dioxide (CO2). Most of FCEVs are equipped with polymer electrolyte fuel cell (PEFC) power system, which is composed of PEFC stacks and auxiliary instruments. PEFC stack has hundreds of membrane electrode assemblies (MEAs) that polymer electrolyte membrane (PEM) is sandwiched by two catalyst layers. PEM has two functions: transporting protons and separating anode and cathode physically and electrically. A catalyst layer is porous structure composed of platinum loaded carbon (Pt/C) nano-particles and ionomer. Ionomer in catalyst layer has also two functions, that is, transporting protons and binding Pt/C nano-particles.

Perfluorosulfonic acid (PFSA) polymer electrolyte, which is a sort of hydropolymer, is often used for PEM. During operation of PEFC power system, humidity in PEFC stack increases and decreases in proportion to output power of PEFC stack. This phenomena is well-known as dry-wet cycle and it causes degradation of PEM and increases hydrogen cross-leakage from anode to cathode. PFSA polymer electrolyte is also used for ionomer in catalyst layer. Size of ionomer in catalyst layer is sub-micron, so it is difficult to detect ionomer degradation in catalyst layer directly. However, from recent researches on PEM, it is possible that dry-wet cycle also accelerate ionomer degradation and break the structure of catalyst layer. In order to clarify the degradation mechanism, nano-scale structural analysis of catalyst layer is required.

According to above background, we analyzed the nano-scale structure of catalyst layer by experimental, computational and microscopic approaches. As experimental approach, mechanical properties and electron resistivity of catalyst layer were measured in order to know connectivity of Pt/C nano-particles. Based on these measured results, we estimated nano-structures of catalyst layers by 3D geometric model analysis. We also observed cross-sectional structures of catalyst layers by electron microscopy in order to validate estimated structure of catalyst layer.

In this research, we analyzed nano-structures of catalyst layers that were prepared on different ionomer / carbon (I/C) weight ratios. Young modulus and electrical conductivity of catalyst layers increase with increasing I/C weight ratio. 3D geometric model analysis based on measured properties of materials and components of MEA suggested that the network-like structure can be formed by connecting small size aggregates of Pt/C nano-particles at higher I/C weight ratio. This result indicates that ionomer in catalyst layer affects not only proton transportability, but also dispersivity and connectivity of Pt/C nano-particles.

Acknowledgement;

This work was funded by ImPACT Program of Council for Science, Technology and Innovation (Cabinet Office, Government of Japan). The authors would like to thank AGC Asahi Glass for providing sample PEMs.

534654 Heterostructured Nanocatalysts for Electrochemical Energy Conversion Reactions

Abstract:

Developing a clean and sustainable energy system is essential for the earth's sustainability. The hydrogen/water energy cycle, a clean and sustainable system, receives of great interest recently. Water splitting and formation are the key reactions in the hydrogen/water energy cycle. Developing active and cost effective nanocatalysts for water splitting and formation is important to water splitting for the production of clean H_2 energy sources and for the production of electricity efficiently. This work will present various nanocatalysts for electric driven water splitting and formation reactions, especially for hydrogen evolution reaction (HER), oxygen evolution reaction (OER), and oxygen reduction reaction (ORR). The M/LDHs, MO/LDHs, and MO/M heterostructured electrocatalysts were prepared by various interfacial engineering techniques. Combining advanced characterization and DFT computational techniques is capable of providing the insight of the heterostructured nanocatalysts for HER, OER and ORR. It was found that the heterostructured interface plays a vital role and provides a highly active catalyst for water splitting and formation. The design strategy for the HER or OER or ORR catalysts can be developed via the understanding of the interfacial properties and reactions.

References

1. 1. K.B. Ibrahim, W.N. Su, M.C. Tsai, S.A. Chala, A.W. Kahsay, M.H. Yeh, H.M. Chen, A.D. Duma, H. Dai, and B.J. Hwang, Nano *Energy*, in press (2018)
2. 2. B.J. Hsieh, M.C. Tsai, C.J. Pan, W.N. Su, J. Rick, J.F. Lee, Y.W. Yang, B.J. Hwang, NPG Asia Materials, 9, e403 (2017)

536609 Metal Nanoparticle Surface Wetting and the Mitigation of Humidification Requirements for Proton Exchange Membrane Fuel Cells

Abstract:

Computational and experimental investigations have previously revealed that the surface chemistry of metal nanoparticles used as fuel cell electrocatalysts can influence both wetting by water as well as the morphology of deposited solid polymer electrolytes. However, the impact of nanoparticle surface chemistry on practical fuel cell operation at high current load is unknown. Mass-transport and ohmic overpotentials are found to dominate performance at high load, hindering the impact of significant recent advances in electrocatalytic activity during practical operation. Prior investigations with operating fuel cells have typically focused on altering the distribution and content of ionomer in the electrode as well as the introduction of hydrophilic agents but often neglect the role of metal nanoparticle surface chemistry. In this investigation, we contrast the electrocatalytic and practical fuel cell performance of commercial carbon-supported Pt nanoparticles to carbon-supported and unsupported Pt and Pt-alloy nanoparticles synthesized via a metal-redox scheme that utilizes stabilizing inorganic ligands to alter the nanoparticle surface chemistry. ^{119}Sn Mossbauer spectroscopy, X-ray Absorption, and X-ray scattering are used to characterize ligand surface chemistry and structure. The catalysts are also characterized with respect to their Pt surface atom coordination using electrochemical stripping of Bi and Ge adatoms. In addition, the surface wettability of the various catalyst components (support, ionomer, metal nanoparticles) is differentiated using dynamic contact angle analysis. We determine that hydrogen/air fuel cell operation at high load correlates both to the surface chemistry of the nanoparticles as well as their mass specific oxygen reduction reaction (ORR) activity obtained from rotating disk electrode (RDE) measurements. For a given surface atom coordination and ORR mass activity, catalysts produced via hydrophilic surface ligands permit the electrodes to operate under substantially drier conditions. Comparable single cell maximum power of 0.5 W/cm2 is realized at 20% inlet gas humidification for nanoparticles synthesized utilizing surface ligands versus 80% when an industrial precipitation approach is used for Pt nanoparticles ~3 nm. A remarkable swap in polarization behavior with humidification is observed. Correlations of fuel cell humidification requirements to alternate nanoparticle morphologies with varying surface atom coordination, inorganic ligand structures, and wetting behavior are described.

530710 Porous Structure Based High Performance Electrocatalysts for Low Temperature Fuel Cells

Abstract:

In this presentation, a simple and scalable method to produce ordered-intermetallic FePt nanotubes by electrospinning will be presented. When tested at cathode catalysts, under the US Department of Energyâ€™s reference condition, the activity of face-centered-tetragonal (fct) FePt NTs surpasses that of commercial Pt/C. In an accelerated degradation test at 1.4 V for 3h, the degradation activity rate of fct-FePt NTs is only 10%, whereas that of commercial Pt/C catalysts is 65%. This approach would provide simple route to support-free intermetallic nanotube structure with superior kinetic activity and higher durability than those of commercial Pt/C catalyst. I will also present the effect of the pore size and doping site position on the single-cell performance of metal-free catalysts using wee-defined ordered mesoporous carbon systems. Well-defined large pore sized mesoporous carbons were synthesized by simple self-assembly of block copolymers with carbon and silicate precursors. The single-cell tests prove that control of pore size and doping-site positon directly affects the cell performance by changing the mass-transport properties and utilizing the doping sites in the catalyst layer. NPOMC with precisely controlled doping sites to be near the large mesopores exhibited a remarkable on-set potential and achieved 70% of the maximum power density of Pt/C. The high performance metal alloy based CO_2 reduction electrocatalysis will be also presented.

538344 Hydrogen Generation Ability of Perovskite and Spinel Redox Materials Via Thermochemical Water Splitting

Abstract:

This study compare the hydrogen generation ability of the perovskite, ABO_3 (e.g A- Sr, Ba, Co, B- Mn, Zr, Fe, Ti) and spinel (AB_2O_4) redox materials via thermochemical water-splitting. Each of these class of materials were synthesized using the surfactant assisted sol-gel technique with alkoxide and salt precursors. As-prepared gels were dried and calcined at different temperatures of 500-700 °C. These powdered materials were analyzed for grain growth, specific surface area, phase composition, and crystalline structure using SEM/EDX, BET surface area analyzer and powdered X-ray diffraction. As-calcined materials were loaded as a packed-bed in the continuous Inconel tubular reactor and 5 thermochemical water-splitting cycles were performed at regeneration temperature of 1100°C and water-splitting temperature of 900-1100°C under the N_2 carrier gas flow rate of 35 SCCM. From the transient profiles obtained, H_2 volume was calculated, and the hydrogen generation ability of the perovskite and spinel redox materials is compared. The solid-state transport properties were also investigated using the electrochemical impedance spectroscopy. The hydrogen generation ability of the perovskite and spinel redox materials and process thermal efficiency will be presented.

532531 Solution Combustion Synthesis of Ni-Pt/CeO_2 nanocomposites for Hydrogen Generation Using Catalytic Decomposition of Hydrous Hydrazine

Abstract:

In search of a liquid hydrogen source for PEM fuel cell vehicles, hydrous hydrazine is a promising candidate with a high hydrogen content of 8 wt% and benign byproduct (nitrogen) via complete decomposition. In this work, the decomposition reaction of hydrous hydrazine is thoroughly investigated using several catalysts. The NiPt/CeO_2 exhibits superior catalytic activity and selectivity towards hydrogen generation, making it a promising catalyst for this reaction. Nonetheless, the preparation of such catalysts often involves complicated processes such as multistep synthesis and long reaction time. In this study, NiPt/CeO_2 catalysts were synthesized simply and rapidly by solution combustion synthesis (SCS) of Ni/CeO_2, followed by impregnation of Pt (impregnation combustion synthesis) using different fuel types (hydrazine, glycine and oxalyldihydrazite) and amounts as well as different Pt loadings on the catalyst. The best synthesis parameters were determined and the results were compared to different impregnation techniques and one-step SCS of NiPt/CeO_2.

Synthesis of Nitrogen and Sulfur Co-doped Graphene on Graphite Foam for enhanced Electrochemical Oxygen Evolution and Phenol Degradation

[1]Xiaomeng Guo, [1]Xiaobin Fan, [1]Guoliang Zhang, [1]Fengbao Zhang, [1]Yang Li, [1]Wenchao Peng*
[1]School of Chemical Engineering and Technology, Tianjin University, Tianjin 300072, China,
wenchao.peng@tju.edu.cn

Fossil fuel depletion and environmental pollution are two serious problems nowadays. Electrochemical splitting of water to generate hydrogen fuel and oxygen gas is an important process for clean energy generation. Electrocatalysis could also be used to enhance the sulfate radical based advanced oxidation process for the environment remediation.

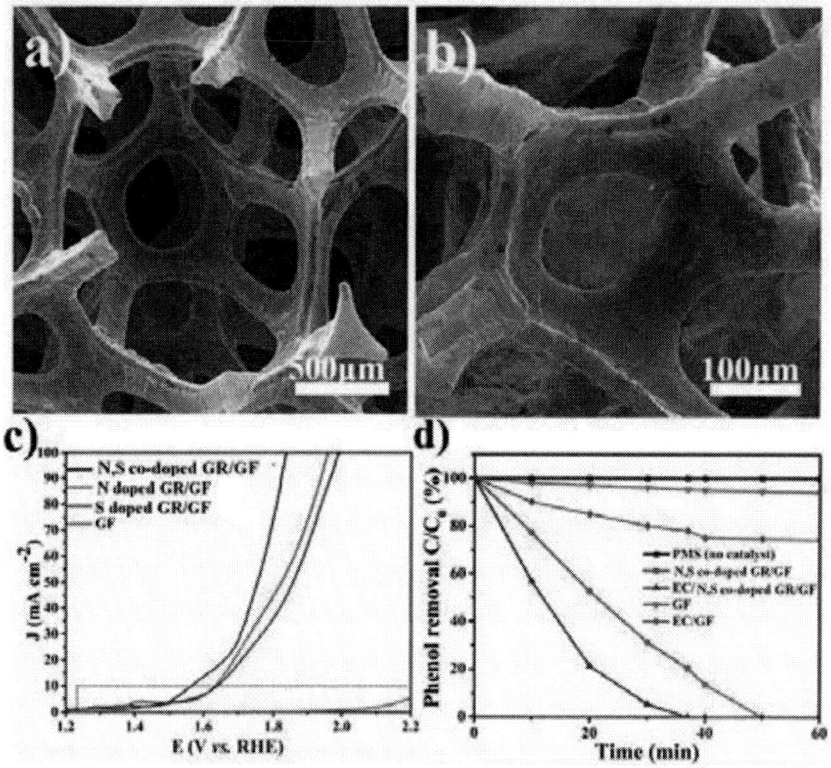

Fig. 1 SEM image of GF (a), N,S co-doped GR/GF (b) and polarization curves obtained with several catalysts as indicated (c); phenol removal under various conditions (d)

In this study, we synthesized N,S co-doped graphene on the surface of porous graphite foam (N,S co-doped GR/GF) by an in situ method. Fig. 1a shows the typical SEM images for GF with porous structure. Fig. 1b shows that the N,S co-doped GR was anchored on the GF surface uniformly, and the structure of the GF was kept unchanged after the doping process. The N,S co-doped GR/GF can be used as a free-standing electrode directly for oxygen evolution reaction (OER) due to the strong mechanical strength and 3D structure of GF. Electrochemical OER activity of the electrodes was then investigated in 1 M KOH solution. In Fig. 1c, S doped GR/GF, N doped GR/GF and N,S co-doped GR/GF show overpotential of 1.63, 1.62, and 1.56 V vs. RHE at a current density of 10 mA cm^{-2} (η_{10}). As shown in Fig. 1d, the N,S co-doped GR/GF shows a good activity for PMS activation after coated with N,S doped GR. The activity for PMS activation could be enhanced further under CV conditions. The improved electrochemical properties of doped GR and porous 3D structure of GF should be important reasons for the activity enhancement. The N,S co-doped GR/GF is therefore a promising free-standing electrode both for water splitting and pollutant degradation.

529419 Development of Cell Reversal Tolerant Anode Catalysts for Automotive Polymer Electrolyte Membrane Fuel Cell

Abstract:

Fuel cell electric vehicles (FCEVs) have been considered as an attractive next-generation automobile due to high energy density, high-efficiency, and zero-emission features. Despite many advantages of FCEV, there are several issues to overcome for the wide penetration into automobile market of FCEV such as stability and cost of fuel cell systems. In particular, during FCEVsâ€™ driving, the various transient conditions cause hydrogen fuel starvation, leading to cell reversal voltage on the anode [1]. To mitigate the fatal degradation of the electrode, materials based approaches including an oxygen evolution reaction (OER) catalyst, oxidation resistant support, functional additive and higher binder loading in the catalyst layer has been pursued [2-4]. Recently, application of OER catalyst has been reporting as an effective way for the reversal tolerant anode (RTA) strategy. In this presentation, the development of non-Pt composition (IrRu) for the anode catalyst having multiple functions for the OER and hydrogen oxidation reaction will be discussed [5]. IrRu supported catalyst (IrRu/C) showed significantly improved stability under cell reversal condition compared to the commercial Pt/C catalyst. And then, the stability and activity of IrRu alloy catalyst towards OER using different carbon supports will be presented. A heteroatom (N or S) doped ordered mesoporous carbon (OMC) supports (N- or S-OMC) were employed as supports for increasing the stability by strong metal-support interaction.

1. C. Qin, J., D. Yang, B. Li, and C Zhang, Catalysts, 6, 197 (2016).
2. R. T. Atanasoski, D. A. Cullen, G. D. Vernstrom, G. M. Haugen, and L. L. Atanasoska, ECS Electrochem. Lett., 2, F25 (2013).
3. B. K. Hong, P. Mandal, J.-G. Oh, and S. Litster, J. Power Sources, 328, 280 (2016).
4. K. H. Lim, W. H. Lee, Y. Jeong, and H. Kim, J. Electrochem. Soc., 164, F1580 (2017).
5. E. You, M. Min, S.-A. Jin, T. Kim, and C. Pak, J. Electrochem. Soc. 165 (6) F3094 (2018).

AUTHOR INDEX

AUTHOR INDEX

AUTHOR INDEX